TRACKS
in the sand

TRACKS
in the sand

a railwayman's war

*Thanks to Biddy, Ann and Wendy
for keeping Tim and Pollie's papers safe*

Also to Iain Logie for answering
many of my railway queries

Extra negative processing by Doug Atfield

Reference books especially useful:
The History of The Corps of Royal Engineers Vo; VI. Chatham. The Institution of Royal Engineers 1952;
The Palestine Campaigns by Lieut-General Sir Archibald P Wavell CMG MC. London Constable and Co Ltd. 3rd ed. 1941;
The Advance of the Egyptian Expeditionary Force July 1917 to October 1918,
compiled by Palestine News. Cairo Government Press and Survey of Egypt 1919;
The Railways of Palestine and Israel by Paul Cotterell. Tourret Publishing 1984;
Hedjaz Railway by R Tourret. Tourret Publishing 1989.

©Jardine Press Ltd 2017
ISBN 978-0-9934779-4-2

Edited by Catherine & James Foster Dodds
Design by Catherine Dodds

All images unless otherwise stated
are from Tim and Pollie's archive.

CONTENTS

Foreword

Introduction

<u>1913 - 15: Brightlingsea</u>
Pollie's memoirs, Tim's letters from home and from army camp.

<u>1915: Gallipoli</u>
Sailing on RMS Aquitania, Suvla, life in the trenches, recovery in Alexandria, home.

<u>1916: El Kubri Light Railway</u>
Sailing on HMT Transylvania, arrival at El Kubri on the Suez Canal, boats, light railway, camels, Indians and Christmas.

<u>1917: The Sinai Military Railway</u>
The advance of the allies from Kantara to Jerusalem.

<u>1918: Jerusalem</u>
The Holy City, Garden of Gethsemane, laying standard gauge track.

<u>1918-19: The Hejaz Railway</u>
The Plains of Sharon, heat and disease, Tul Keram, Haifa, Samakh, demobilization, Zagazig and home.

FOREWORD

Tim Foster, a Railwayman.

The hundred year old letters, photographs and memories that make up this book have been passed down through my family. They have been lovingly kept in a large brown case containing cigar boxes full of tiny negatives, faded contact prints with notes on the back, railway timetables and documents, tickets, newspapers and cuttings in scrapbooks and other sorts of ephemera. Making sense of all this material has for me and my wife taken many years. The letters document the courtship of my grandparents, Tim and Pollie Foster, and their six year engagement separated by the First World War.

Tim first saw action in Gallipoli and was very lucky to survive. In his next tour of duty in Egypt and Palestine he took a pocket camera with him and took a keen interest in recording everything around him. He developed the film himself and made contact prints using the sun, sending some photographs home with his letters. We have used his words from the back of these photographs as captions, and combined with his letters to Pollie we have been able to piece together his part in the war, a war that was largely won by its use of railways.

It has been wonderful getting to know my larger-than-life grandfather, a self-educated, practical, poetry-reading socialist and railwayman, who died a year before I was born. Deciphering Tim's faint pencil-written letters and his faded photographs has been challenging and consuming, and piecing together their story against the backdrop of the war has brought them back to life for us. We hope that you also find their tracks in the sands of time equally fascinating.

James Foster Dodds & Catherine Dodds

Millie, Father, Fred, Jess, Timothy, Mother, Lydia.

INTRODUCTION

Timothy Charles Foster was born in Ipswich on the 13th January, 1888, the son of Joseph F Foster and Emma Haddock. Joseph was an iron machinist and trade unionist who worked for Ransoms Sims and Jefferies of Ipswich, builders of traction engines, agricultural machinery and the first railway engines in China. Tim was the fourth of five children, with three sisters and one brother.

In 1901 the Fosters moved to Hadleigh in Suffolk when Joseph took up the job as manager of the gas works. Tim, aged 13, started working for Great Eastern Railways from his home town station. Here he met his life-long friend Alfred Blundell, whom Tim encouraged to give up his job on the railway and go to art school. Alfred subsequently became a well respected artist. After Hadleigh Tim worked at Manningtree, Eye, Rayne, Walton-on-the-Naze, and Ipswich stations.

At the age of 14, Tim was working nights at Manningtree station in charge of the telegraph when he fell asleep and sleepwalked. He was found the next morning asleep on someone's doorstep the other side of the town.

While he was working at Ipswich, on the 1st October 1911 there was a movement amongst the clerks in the service of GER for better conditions of service. It was stated by the clerks that the ordinary maximum of clerks on the GER was £67/10s compared with £110 on the Midlands Railway. Tim (aged 22) presided at the Ipswich meeting in the Cooperative Hall. A resolution to form a branch of the Railway Clerks' Association was unanimously passed. By December Tim was referred to as the secretary of the Ipswich branch of the Railway Clerks' Association and later as representing the Commercial Dept, District No. 3 Brightlingsea.

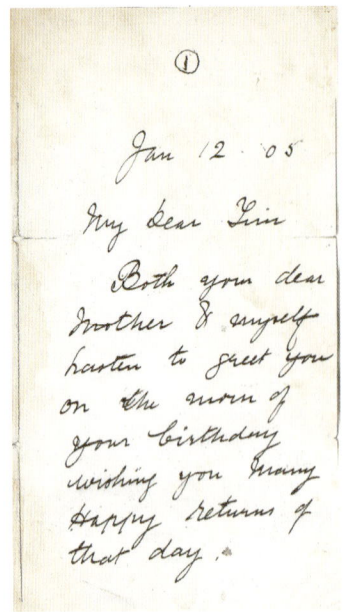

Letter from Joseph and Emma to Tim on his 17th birthday.

An alarming postcard Tim sent to his parents. He was 17, and had a mischievous sense of humour.

Jan 12 05

My Dear Tim

Both your dear Mother and myself hasten to greet you on the morn of your birthday wishing you many happy returns of that day.

This is a "measuring" day for a young man, when he should look at himself and see if his mind, his intellect, is growing with his body and also see if he is developing a true good character.

We both of us believe you are doing your best to see that all these points of your nature are truly and lastingly improving.

We have sent to Lydia, and Millie and Jessie too will join us in sending you a small present on your birthday. Kindest love and a birthday kiss from dear Mother and myself with best wishes for your birthday

*From Your loving
Mother and Father*

I wonder what will be the result of the Ipswich election today.

Tim seated on right at station, possibly Rayne, Essex.

From Ipswich station Tim moved to Brightlingsea in 1912, which is where he met his future wife Pollie. Pollie (aged 16) and Tim (aged 26) became secretly engaged just before Tim enlisted on the 7th September 1914 in the 1/5th Suffolk Regiment. Before embarking to Gallipoli in July 1915 they became officially engaged.

After basic training in Thetford the 1/5th Suffolks became part of the Suvla Bay landings in Gallipoli on the 6th August 1915, as part of the Mediterranean Expeditionary Force (his graphic account of his part in this disastrous campaign was published in the Brightlingsea Times October 1915). It was very difficult to keep the troops supplied with clean water as they were virtually pinned down by the Turkish guns, and Tim like so many men succumbed to typhoid. He was put in the mortuary tent, being told that he would be dead by the morning, but he was rescued by his brother Fred, a medical officer, who also went down with typhoid. Fred was later mentioned in dispatches as keeping the men's feet in good order in the advance across the Sinai desert. Tim spent October 1915 in hospital in Alexandria, then returned to England on HMHS Aquitania to a hospital in Birmingham, and then home leave.

Departing for Egypt on the 26th February 1916 Tim was part of what would become the "Egyptian Expeditionary Force". This time Tim took a pocket camera with him keen to record all around him, developing the photographs there and sending pictures back to Pollie with a flood of passionate letters and postcards. In March 1916 he was at El Kubri "C" Subsection, No.1 Section, Canal Defences, where he was in charge of the light railway from the No. 2 bridgehead on the Suez canal to the trenches in the desert. By December 1916 he had been transferred to the Royal Engineers, 276th Light Railways Company in preparation for the running of the Sinai Military Railway.

Unlike the relative stalemate of the war in France and the failure of the Gallipoli landings, the war across the Sinai Desert was a fast-moving campaign. The Turks where building a railway down through the Sinai desert to supply their troops who wanted to capture the Suez Canal and cut off England from India and Australia. In 1915 the Turks had advanced to the canal and sunk a few ships, but were unable to cross or keep their troops adequately supplied across the desert, hence the need for a railway. Troops from England and the Empire (Egypt, Canada, Australia, New Zealand, Indian and Gurkhas etc) were dug in along the Turkish side of the canal as a defence force. Then began the offensive, building a railway up the coast under the protection of naval guns to link up with the Turkish railway (a different gauge track). The first to build the railway across the Sinai desert would win.

Tim became part of the building of this military railway across the desert, working at various stations along its advance. When Jerusalem was captured he became the first English station master, and when the Turkish track became fully converted to the English standard gauge, Tim was in charge of the first train to run from Cairo to Jerusalem, a tank engine commandeered from the Inland Waterways and Dockland Railways, Liverpool. He then became station master at Tul Keram, Haifa and Samakh as the English force advanced north, and worked for the Egyptian State Railway when unrest broke out in Egypt after the war ended.

Tim found Jerusalem a real delight after the Sinai Desert, and he wanted Pollie to join him out there after the war, but she would not. He eventually returned home seven months after the war had ended and married her (now aged 21) on the 17th November 1919 in the Swedenborgian Church in Brightlingsea.

Pollie was the youngest of seven children, the daughter of William & Emily Pannell. Her siblings were, in order: Emily, William, Tom, Mary, Dolly and Fred. Her father William was self-made business man, who ran many businesses in Brightlingsea, from grocery store and sailmaker's to ships and farms, and served on the Urban District Council. Pollie's elder brothers had been educated at private school and disapproved of her marriage to a lowly socialist railway clerk. William however wholeheartedly approved, having started work himself aged 12, he perhaps saw a kindred spirit in Tim.

After their marriage Tim and Pollie lived with Joseph and Emma in Hadleigh. Pollie would watch for Tim's return from Ipswich by train from a small square window in the otherwise blank back brick wall of their house. The house and window can still be found to this day in Bridge Street Hadleigh. Tim had resumed duty with GER in July 1919 and served in the district office Ipswich until August 1920, when he moved to Brightlingsea as a goods clerk.

In 1922 Tim was one of the British Legion's nominees for the Brightlingsea Urban District Council on the 3rd April, and was elected (and served for three years) to bring about the job creation project for the unemployed. The West Marsh improvement scheme would employ 30 men for 107 days at a cost of £7,245 of which the Government would give £6,000. This brought about the building of the Brightlingsea promenade. Whilst a councilor he exposed corruption by the Brightlingsea Gravel Co. in April 1923, and his father-in-law (who was not involved in the scandal) felt he should resign as chairman on the Council. Tim had to wait fourteen years to became a station master again, at Brightlingsea for the LNER in 1933. He subsequently was station master in Hatfield Peverel 1936, Wickford 1939, and Harwich during the second world war.

Election flier 1922.
Tim's medals: the 1914-15 Star, the British War Medal, Victory Medal and ID tags.

JERUSALEM'S FIRST BRITISH STATIONMASTER

Ipswich Railwayman's Promotion

Mr. T. C. Foster, stationmaster at Brightlingsea, has been promoted stationmaster at Hatfield Peverel. Mr. T. C. Foster, who is a native of Ipswich, started his railway duties in the parcels department at Ipswich Station, being subsequently transferred to Brightlingsea in 1912, where he took over the duties of goods clerk. Enlisting in early September, 1914, in the 5th Battalion Suffolk Regiment, he took part in the Suvla Bay landing at Gallipoli, and afterwards saw service in Egypt. He was stationmaster at various stations in Palestine, including Jerusalem (at which, incidentally, he was the first British stationmaster), Haifa, Samakh and Tulkeram. During the riots in Egypt after the war he was on the Egyptian State Railway. Resuming civilian life in 1919, Mr. Foster restarted railway duties at Ipswich Station, but in the following year (August, 1920) he returned to Brightlingsea. In 1933 he was appointed stationmaster there, and it is unique that he is the only stationmaster to have been promoted from that station. It is worthy of note that during the time he has been at Brightlingsea the traffic, both passenger and fish, has increased tenfold. Mr. Foster served a term of office as a member of Brightlingsea Urban District Council and proved a very active member. He moved the resolutions whereby both the first and second schemes in connection with the West Marsh pleasure grounds were brought into operation. He married in November 1919, the daughter of Mr William Pannell, a past County Councillor and chairman of Brightlingsea Urban District Council.

SEQUEL TO COLLISION

STATIONMASTER AT JERUSALEM

Harwich Official's Unusual Career

Mr. T. C. Foster, who has recently taken up duties at Harwich as stationmaster, has had an unusually varied and interesting career, having had the distinction of taking the first train to Jerusalem where for a period he was station-master.

Mr. Foster came to Harwich from Wickford, having been at Brightlingsea and Hatfield Peverel previously. He started his railway career 40 years ago at Hadleigh and saw service at Manningtree, Eye, Rayne, Walton, Ipswich, and Brightlingsea before the Great War.

He joined the 5th Suffolks in 1914, and saw service at the Dardanelles, and was present at the landing at Suvla Bay. He was invalided home but recovered and returned to Egypt, where he was transferred to the Railway Operating Division. He was in charge of the railway at Suez for a time and went out with the railway across the desert and on to Jerusalem. He took the first train, comprising an engine pushing a truck into Jerusalem, where he subsequently became stationmaster. He was in charge of the first standard guage train to run through from Cairo to Jerusalem.

From Jerusalem Mr. Foster went to Haifa and Samakh and also served for a time as a stationmaster on the Egyptian State Railway.

After the war he returned to Brightlingsea, where he remained until 1936.

He grew fruit and veg from the station garden, and kept goats, chickens and bees, providing for his doting family — Pollie and their three girls Bridget, Ann and Wendy (my mother). His last station was at Marks Tey until 1951 when he retired back to Brightlingsea. Tim died 22st October 1956 aged 68.

So now in their own words is the story of Tim and Pollie's courtship and a railwayman's part in the war.

Pollie Pannell and Olive Eagle.

32 Victoria Place, Brightlingsea (built by William Pannell). Pollie's bedroom was in the top of the turret.

1913-15: BRIGHTLINGSEA

<u>Pollie's Memoirs, written for her daughters.</u>

1913: The last year of my school days! I was not clever or even very interested in lessons, always glad when holiday time came, and during the last few months of school I became much more interested in Tim Foster, clerk at the Railway Station.

Before going to school in the mornings I used to walk to the station with Olive Eagle when she went off to Colchester by train to Colchester High School, I would also meet her off the train at 5.15.

One morning I happened to glance towards the ticket office window, I saw a dark curly-haired young man bent over the desk. I exclaimed to Olive, "Did you see that head with masses of shiny black curls?" Next morning I peeped again, this time I saw his face, and thought "What a handsome young man!" Little did I think then that he was to become your father, Bid, Ann and Wendy!

From that day onwards he was the main interest in my life!

I soon found out that Tim lodged at Jenny Bagley's in Queen St, opposite where Olive lived, going down to the station to book the first train out at 7.30. This was the time I was getting up so saw him going past the house.

Breakfast always on the dot at 8am. I was supposed to do my music practice between 7.30 and 8 o'clock. At 8.30 I would help my mother make the beds, very carefully, mother telling me "A well made bed brings a good husband!" Then off I would go to see Olive off on the train at 8.40, always glancing towards the booking office! After a few weeks I discovered that Tim left the station as soon as the train had gone, to go to breakfast at Queen St. Needless to say I would loiter on my way home, then one day I actually got a smile from him, usually so very serious, that made my day.

Two mornings in the week I went to school early to have a music lesson before school opened. It was on one of these mornings with music under my arm I was walking up Station Rd after seeing Olive off when the serious young man caught me up and in passing said "Good morning Miss Pannell!" This more than made my day.

It was a lovely hot sunny day in May (7th). Such a day when it was hard to be shut up in school, the garden at my home gay with flowers, the trees just at their best, copper beach and weeping ash, garden chairs with bright red cushions. Always in May I can see the garden as it was that day and wish I had a coloured photo of it.

On this lovely May day I didn't know how to wait until Olive got home to tell her Tim had actually spoken to me. My music lesson didn't go too well that morning, Miss C saying "What is the matter today Pollie? Nothing but mistakes!"

Every year on the 7th of May I am transported to that walk up Station Rd, sometimes it is just such an early heat-wave day. Sometimes cold or wet, but always I see it as it was in 1913.

It was a little time before I found out Tim's name. A Belgian girl who came to school and in my class always referred to him as "Feutre Vert" meaning "Green Hat". He always wore a green felt hat and neat suit, smart and immaculate! Can you believe it, girls? You who remember him as so careless of his appearance, this was before the days when he had so many hobbies and no time to trouble about his clothes.

All through these summer days Tim would have a word or two to say to me each morning, until eventually we boldly walked together up Station Rd.

This summer Mary and George were married. I was bridesmaid again, wearing a blue dress, embroidered with pink, made by Doll. I also wore a huge cartwheel straw hat with blue and pink forget-me-nots around the crown. I told Tim I would stand in the front door way at Victoria Place as he was passing he could see me in my wedding get up.

One day someone told my mother that I was "chasing the railway clerk" and often seen talking to him outside the Reading Room, or walking up Station Rd, going towards school. Mother was very disturbed because the only clerk at the station that she knew of was a middle-aged peculiar man, and she was horrified! When she spoke to me about this I said, "you wait until Tim Foster comes past. I will point him out to you, and you will see he isn't the one you think." And so one day we met at the gate, I introduced them. Mother liked the look of him and was very relieved that it wasn't the other chap. No more was said.

During Tim's dinner hour he would call in at the Reading Room where he was a member. This was usually at the time I was going to afternoon school. He would walk down Duke St as far as the cut then go through by the National School to the station.

One day I played truant from school. Tim and I walked towards the farm down East End Green. On our way back, to my horror, who should be coming along the road but my father. We quickly got over a gate and hid behind the hedge in a field until we thought he would pass. We saw him look over the gate at the far corner of the field, whether he saw us or not I never knew, perhaps he turned a blind eye. I didn't feel very happy or comfortable, sitting at the tea table that afternoon, nothing was said!

At the end of this year St Hilda's was to close down. It was suggested that I should have another year at Colchester. I said "NO!" When St Hilda's finished I would be glad to pack up my school days.

Soon a voluntary job was found for me. In January 1914 Dorothy Lambert was born, and instead of school for me, at 9 o'clock I was expected to make myself useful to Mary, this became my unpaid job until I was married! Every morning after seeing Olive off I would walk up with Tim to Queen St on my way to Regent Rd. First job was to wash up breakfast things while Mary bathed the baby. I would then give Dorothy her bottle. When settled in her pram I would take her for a walk, do shopping. Whenever George had a meeting in the evenings I had to keep Mary company. How fed up I got with this over the years.

Some evenings I would go over to see Winnie Whitaker. Sometimes we would call in at the station to see Tim in his office. One evening a boy called Wally Aldous was there. He was staying at Bagley's for a holiday with his parents. Winnie always an excitable young miss soon got talking to Wally, he dared her to smoke a huge cigar that he'd got. This Winnie did, right to the end. Wally was killed early part of 1914 -18 war.

Life went on all happy and peaceful until 4th August 1914 when war was declared. Everyone was horrified, but hopeful it would soon be over. Life quickly changed for everyone, men were enlisting and getting into uniform. Ration books were issued.

On 21st August I had my first (unofficial) engagement ring! How thrilled I was, Tim had been to Jarvis the jeweller. Mrs Jarvis helped him to choose

Mother, Doll, Pollie and Fred Pannell (Fred lied about his age and enlisted a year early. He was a sharp shooter at the Somme).

a selection to bring for me to pick which I liked best. Sitting in a 1st class railway carriage, stationary coach, I chose my ring. It was difficult to decide as all were so dainty and pretty. The one I decided on was set with pearls and turquoise. This I wore round my neck on a ribbon under my dress, but on my finger when out with Tim. A few years later I lost it on one of my visits to Moverons. I always thought I must have pulled it off with my glove.

Those summer evenings we used to walk along the line towards the bridge, this was before West Marsh was developed. It was a lovely natural marsh, with sea daisies and sea lavender. You would be amused had you seen us sitting on the grass or sand near the bridge, Tim reading poetry to me. All the young couples who walk along the line 50 years later would be highly amused, they carry pocket radios and listen to pop songs!

One day Tim's sister Jess and her boyfriend Len Oxborrow turned up. I was thrilled to meet them, fell in love with them both. Jess a lively pretty girl, Len very good looking, they were amused to see me with Tim, didn't think there was anything serious between us. Later Mr and Mrs Foster came to B'sea. I joined them on the Hard with Tim and we went over the ferry, walking along the beach Tim's mother laughed and teased me, telling me I would change my mind many times before I would settle down with a proper young man, that I was far too young for her Timmy. I felt rather crest-fallen, but assured her I wouldn't change my mind.

This summer we were playing tennis at Cater's. Tim was also a member of the town Cricket Club. I used to watch the matches on the Red Barn farm Meadow. A nice walk through the cornfields to get there (now a council estate). A member too of the Reading Room, he and his friend Arthur Portass (schoolmaster) formed a debating society, both of them very eloquent speakers, for years their speeches and ideas were spoken of and remembered. Portass thought I was good enough for Tim.

On the 7th September 1914, all was changed: Tim joined up. He went to Colchester to enlist in the same regiment as his brother Fred who had been in the army a few weeks. I said goodbye to him early in the morning and by afternoon post I got a postcard from Colchester "I am now a private in the 5th Suffolk Regt. Tim."

I was downhearted and proud. Downhearted knowing I would no longer see him every day, proud because he had joined up, when young men were getting white feathers handed to them by unkind people when they hung back from joining up. I didn't want Tim to be handed a feather! I had my ring and was engaged to a soldier, that was a great comfort.

The following Sunday he came down to B'sea on 2 hours leave in his new uniform and was able to do so fairly often while stationed at Wivenhoe and Colchester. Once, mother and I went to Colchester. Tim was able to meet us for tea. On our way home, we were stranded on Hythe Station for hours in the black out, while troop trains kept rushing through. We didn't do that again. Tim could come when he could by road on his cycle, would cut it fine going back to camp before roll call, if he was late his pals would cover up for him.

By this time Tim was invited to Victoria Place when on leave. Brother Tom didn't approve of this... snob! Fred joined the Essex Yeomanry and was soon in France, little did we or anyone else think it would be such a long horrifying war. Everyone believed it would be over in a little while.

The time came when Tim expected his embarkation leave and on the 16th of July 1915 he asked my mother and father if we could be officially engaged. They agreed and we went off to Ipswich to buy the ring, my mother asking Tim not to buy a big, flashy or expensive one, something small and modest, hence the very little one we chose. We then went on to Hadleigh and stayed overnight. What a welcome we had, and how surprised I was to see Tim and his father kiss each other on arrival and departure. I had never seen this happen when a son was grown up.

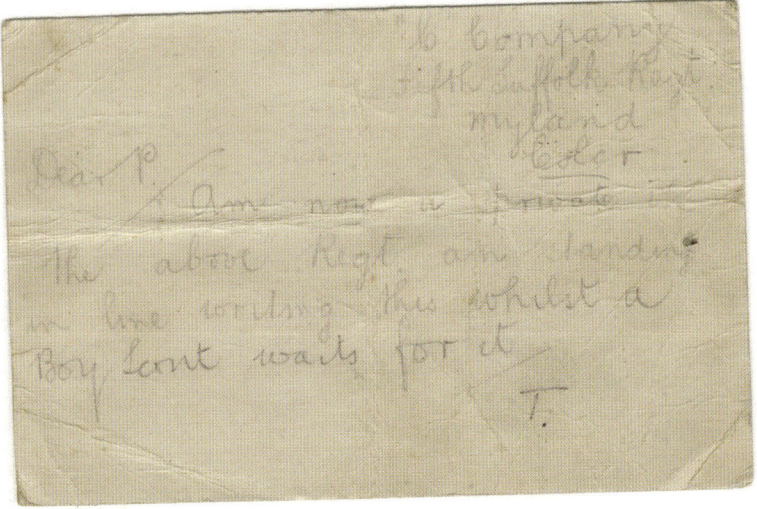

"Dear P. Am now a private in the above Regt. Am standing in line writing this whilst a Boy Scout waits for it. T"

Sunday Eve

Dearest

   Your letter found me in bed. Jessie brought it to me, it came as a pleasant surprise for I feared you would address it to Colchester. It was sweet and loving. How I should loved to have stolen to you before you went to sleep on Friday night, but of such bliss I cannot speak or write, that belongs to the time when I close my eyes before sleeping. Is it not then that we come close to each other My Little Pal, My Love.
   Yes, Blundell is here and Oxborrow too. B is playing the piano and O is before the fire with J.
   My darling how I should love to be with you tonight, to have you all on my own. I know I am but a poor lover and often dull and wearisome, nevertheless I want you.
   Blundell has just chirped in requesting me to send from him a few xxxx and Jessie wishes me to send her love.
   I am due back in camp again tomorrow night.

   Fondest love and many kisses to you darling,
   Your Man.

Regimental wallet containing Pollie's photo.

Wednesday Evening

Dearest

   Did you see the sun set today? It was beautiful. At about 4o/c we were in a meadow nearby when an aeroplane passed over high up in a beautiful pale blue sky. We watched it getting smaller and smaller as it went westward until it disappeared into the setting sun. It was a beautiful sight. It seems as though it passed from this world altogether into another beyond the sunset. Anything leading or passing from the known to unknown, from reality to fancy, appeals to me. We stand in the present day — the known, and look forward to the future — the unknown. And we reach forward and in doing so realise step by step that which at one time existed only in our fancy — or in our dreams. All this because of a beautiful sunset. You'll be going to sleep if I don't leave off.

   My dear girl, how can I express to you my pleasure when I saw that which you slipped into my pocket last night. It's a treasure which I shall always carry with me and hold dear. It's a very good idea and very thoughtful of you. Darling I should love to be with you now, that I might take you in my arms and kiss you. I am very happy you have made me so. Dearest you gave me a most tappy tune last evening. I must run to catch the post or you won't get this in the morning.

   Yours,
   Tim

Wednesday evening

Dearest

We went past Hall Farm Ardleigh today. Not many of us, only about half a score. Didn't see anything of the Pannell family.

This evening I saw some armoured motor cars, there were some motor cyclists too. On Friday we have a big field day at Elmstead and Alresford. I expect there will be some 5 or 6 thousand of us there sham fighting. There are some wounded at Ardleigh. We had a chat with one of them he was struck by a shell he told us.

I made plasmon for supper last night. Unfortunately I let it burn a bit but still it was good. I think I shall have to get an enamel saucepan. Lack of convenience and utensils made it impossible for me to proceed in accordance with the directions and so I went like this:

Taking the only saucepan available I poured a cup of water in it and made it luke warm. Then the plasmon was introduced. It took quite a lot to make it look anything like "porridge". It was so thin, so I kept putting more and more in until at last I thought I'd better stop and save some for another evening. After the plasmon a little more water was added, making the saucepan about 3 parts full — 2 cups of it there looked — then it was placed upon the fire again and before I could get my watch out or find a spoon to stir with the contents were pushing the lid off and running down the sides of the saucepan into the fire. So I had to take it off whilst I ran and begged the loan of a spoon next door. Then it was put on the fire again without the lid. It soon boiled up again and ran over, so I dropped the spoon in, intending to stir it according to the instructions, burnt my fingers in getting it out, dropped the spoon in the ashes of the fireplace, and burnt my other hand on the saucepan handle. All the time the plasmon was boiling over into the fire and I could not stop it, and it was very thick too. Of course I'd put too much in, however it was very nice, but next time I think I shall use less of the plasmon, and use sugar instead of salt.

In the absence of a basin I had to eat it out of the saucepan. I hope you don't mind.

It would be very nice if instead of writing you I could slip out and meet you up the Lower Park Rd as we used in days gone by. We had some happy times didn't we?

Good Night Sweetheart,
Yours,
Timothy

What do you think? It was about half past ten last night, we were all laid down for the night, lights were out and I had just tucked you away in your proper place when a wave of restlessness swept over the room, and before I could secure it my bed and blankets were swept away from me. There was great confusion, 8 pairs of legs, 8 bags of straw, 24 blankets, 8 heads, 8 kit bags, water bottles, flags and rifles were all in a heap. Somewhere amongst them was a little leather case containing a photo, there was a fierce struggle to get on top, those beneath battled to get out, and when they did those on top went under. For a full hour this went on. Never did I see (or rather feel for we were in complete darkness) such confusion — blankets, beds, kits, legs and heads were flying about like a whirlpool. It was nigh on midnight before the struggle subsided. Then came a sorting out of persons and belongings. Vainly I fumbled about for the case. At last an exclamation of surprise announced its finding, like a flash grasped it, but it was no good trying to conceal it, everyone demanded to see it and what was in it. Electric torches flashed out and you were admired by the whole room, the remarks passed made me feel quite proud. Charlie identified you and I believe when they eventually closed their eyes most of them were half dreaming — thinking of you whilst I held you more closely than ever.

Now my love I will conclude with fondest love,
Yours,
Tim

[scrap]
Really if you keep my letters in your old school bag or anywhere where other eyes than yours may read I must write according. It would not matter if they were well written, as really good love letters should be. Some are even good enough to be bound together in books affording good literature to poor scribes. But mine, poor, badly written, full of errors, and weak expressions of my love to you will not bear being read by anyone bar the one to whom I look for forgiveness for all my shortcomings...

6.11.14
Colchester

Dearest

Oh no I'm not going to the front this week. Some of our chaps went this morning. Had you been with me half an hour ago you would have been marching behind them, then our ways parted with three mighty cheers. They marched on to the railway station and we turned off to the golf links. They happened to be departing at the same time as we (Signallers) were marching out, so we brought up the rear and kept them company part of the way along the Mile End Rd... There was a long succession of hearty cheers, someone blared out "Auld Lang Syne" on a trumpet or bugle. There were some moist eyes and sad faces too.

No I'm not going this week or next. In fact I don't know that I'm going at all. One battalion only of our regiment has gone — that's the 4th — we are the 5th. Many of our chaps in the 5th Battalion are very disappointed they're not off too.

It's 10.45am and I'm out in the country flag signalling. It's a beautiful autumn morning, there are some lovely trees around covered with leaves of the most beautiful shades. Every now and again a leaf flutters down from the tree beneath which I write and gently drops to the grass all wet with the fog of the early morning.

Good night my love, I hope it will be fine one Sunday.
Yours,
T X

2394 L/C TC Foster
Signal Section
A. Company
1/5th Suffolk Regiment
163rd Brigade
54th Division

1/5th Suffolks. Tim standing in the middle.

13.11.14

Dearest

Many thanks for letters. Our marching orders were cancelled as suddenly as they were given.
I will, weather permitting, be at Ardleigh Church by 5.30pm (Saty)
Excuse haste,
Yours,
Tim

Hadleigh
24.6.15

My Dear Poll

Jessie was very pleased with the heather and wore some of it in her buttonhole. Her husband wore some too, it was very kind of you they said, and were glad to hear you were well. Mother too was pleased and said "Why of course you must go to B'sea!". They half expected to see you and those guests who knew of your existence enquired of you at the breakfast — which by the way came about dinner time. Someone not in the know spoke of me as the "bachelor uncle" but Jessie said — "Oh no! We've had a telegram which says differently" and so your telegram with others went round. Jessie was very delighted to get a telegram from you and everyone wanted to know who "Poll" was, so you came into the limelight. The whole affair went off remarkably well and the happy couple left here about an hour ago (5o/c) for their home in Ipswich in a Motor Car with a pair of white satin slippers hanging behind.

Both of them were very happy indeed and in high spirits. The Bride wore a lovely buttonhole in which was some white heather from the Best Man's fiancee. The best man was in form looking fresh having arrived by the 10o/c train from B'sea where he had been spending a few happy hours with his sweetheart. He was happy too for his love had met him that morning with a sweet smile which made his heart glad.

Yes, dear, your face this morning was sweet and happy and the sight of you made me glad.

Everyone sends love to you,
Tim

Pollie's address book:
Tim's many different addresses starting at home, then with the Suffolk Regiment, and finally with the Royal Engineers in the ROD (Railway Operating Division).

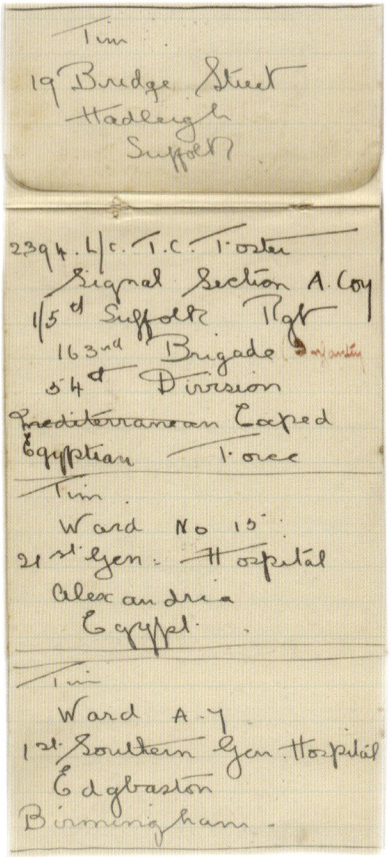

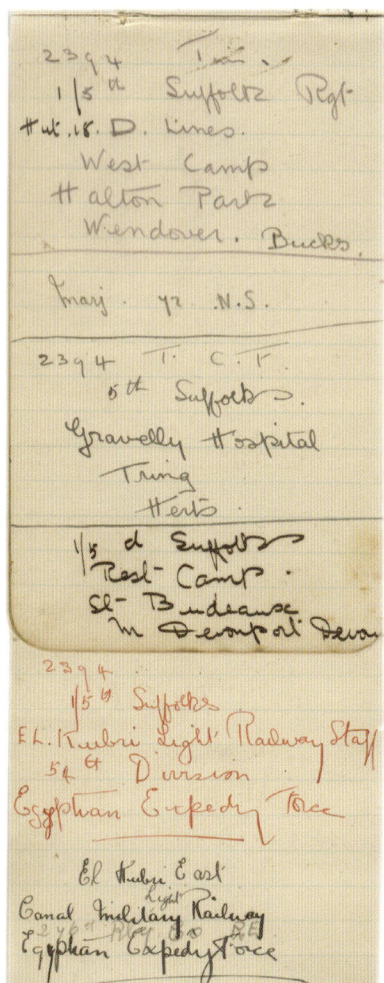

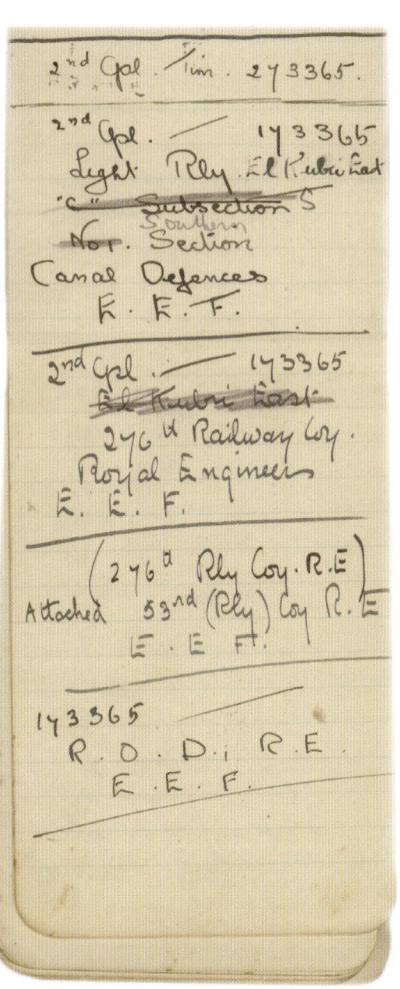

Tim used whatever paper he could lay his hands on to write to Pollie. The YMCA provided writing paper for the troops.

The Times.

LONDON, FRIDAY, JANUARY 7, 1916.

ANZAC AND SUVLA.

FULL TEXT OF SIR IAN HAMILTON'S DISPATCH.

1915: GALLIPOLI

The campaign in Gallipoli was an attempt to capture the Ottoman capital of Constantinople. On 30th July 1915 the 1/5th Suffolks took the *Aquitania* from Liverpool along with seven or eight thousand men. They were escorted by two destroyers but by the morning a severe gale meant the escorts had to return to Liverpool and the *Aquitania* continued alone, avoiding German submarines by sweeping wide out into the Atlantic. They passed Gibraltar on the 2nd August and Malta on the 4th. They arrived at Mudros on Lemnos in the Aegean Sea on the 6th August.

On the 10th August they were ferried to Suvla Bay in Gallipoli to join the Anzacs to help fight the Turkish Army. The landing and subsequent action was mismanaged and the Allied forces failed to gain much ground inland. By 15th August they had advanced only 1,500 yards under heavy fire. Inside 72 hours, 11 officers and 178 other ranks of the 1/5th were killed or wounded, and stalemate ensued.

Tim's last letter from the trenches was 20th September. Due to the unsanitary conditions, and like many other soldiers, he caught typhoid, or enteric fever, and was in hospital by 11th October. His next letter to Pollie was on the 31st October, from a military hospital in Alexandria. He came home on the *Aquitania* for Christmas. The rest of his battalion began evacuations on 11th December. The British Empire and French forces lost around 44,000 men in the failed campaign.

From The Times, 7th Jan 1916

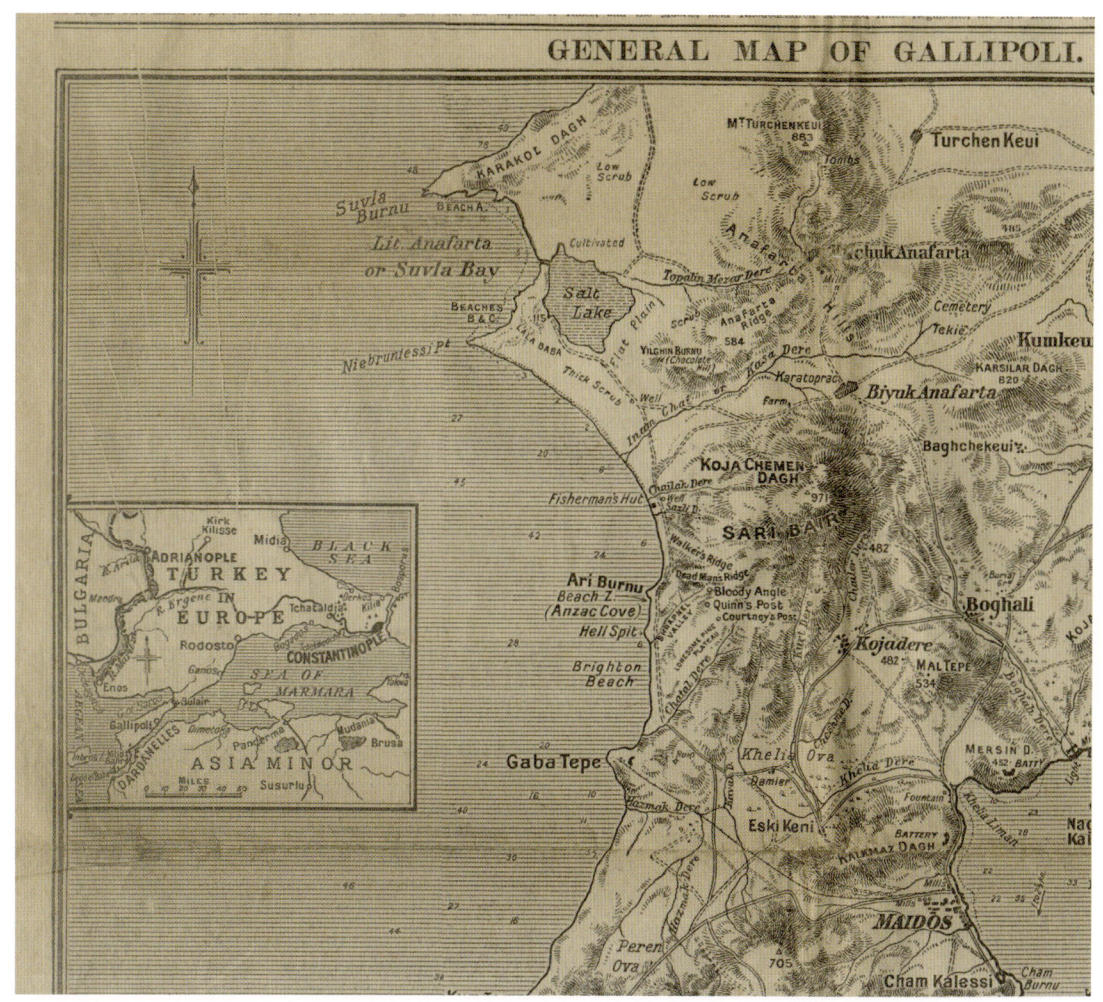

Soon after the embarkation leave the 5th Suffolks sailed east and landed at the Dardanelles, the beginning of the dreadful time at Gallipoli and Suvla Bay. I won't dwell too long over this, the dreadful food, precious little water, the dreadful horrors and hardships, all for soldiers' pay 1/– per day. Only those who were there know how dreadful it all was. Before long those who were not killed or wounded were terribly ill with dysentery and fevers, heat, lack of water, and this a terrible torment.

Fred was in France, going through similar horrors. Trench warfare too terrible for words: frost-bite added to wounds and illness, eating, sleeping, fighting, living and dying in those dreadful rat-infested trenches and dug-outs.

One late evening the front door bell rang at Victoria Place. I went to open it after putting lights off and pulling aside the blackout curtains. There on the doorstep stood what appeared to be a gruesome lump of khaki, loaded with kit and rifle etc. It was Fred [Pannell] home on leave, straight from the trenches, poor boy. Mother nearly collapsed with shock! He was in a filthy state, covered with mud, unshaven and tired out with lack of sleep. Worst of all, alive with lice! He had to undress in the scullery, throwing his clothes outside. What heaven it was for him to get into a hot disinfectant bath. Imagine how we all felt when he had to return to France and more horrors after a few days leave. Letters and field cards were few. Fred had little to tell.

I was fortunate. Tim was a good letter writer, and managed to write as often as possible. Over the 4-5 years he was away — mostly all the time in Palestine – I must have received many hundreds. Sometimes letters were held up, some lost when boats carrying mail were torpedoed, and some would come a whole batch together. When this happened my father used to tease me, wanting to keep some back to hand out to me on the days there were no letters. He never did. I would go off to my room at the top of the house to read then sit down to write to him.

Aquitania, as a troop ship in dazzle paint scheme in 1915. She became a hospital ship after 3 transport voyages.

29.7.15
10pm

Dearest

It was 5.30 when we marched off to the station at Watford, and your letter was brought on by a later train and has just been handed to me. Needless to say I was delighted to get it.

I was very pleased to find you writing so cheerfully. I've just received a letter from home and it struck me that Mother might be somewhat downhearted, for her only two sons are leaving at once. If you would drop her a letter occasionally, particularly just now, I'm sure it would cheer her up a lot and at the same time would be doing me a great kindness, that is of course if you felt so inclined. Don't be disappointed or surprised if you had no reply or if it were but brief for as I have told you before she is not much of a letter writer. Then again you will hear from me probably more than she will and it would be doing me a kindness in that way by letting her know how it goes with me.

I'm not exactly sure what I may write and what I may not. I don't suppose the sensor would object to anything in the "love" line, though I have doubts as to what he would do if I told you that we are sailing on the ... from ... I know we are not allowed to give the date of departure or our destination, but that doesn't matter I can't tell you any more than you know about that. This ship is a ... and it's very difficult to find one's way about, in fact we've spent the greater part of our time at present on finding ourselves. Charlie and myself have got a very comfortable 2-berth cabin to ourselves and we don't mind how long the voyage lasts. I hope to get this letter passed by the censor before we leave.

To you I owe my happy state of mind at this time, to you I owe much more beyond this, to you I gladly give all that I have.

Yours most affectionately,
Tim

30.7.15

We live like dukes with one or two slight exceptions, and I no longer sit on the floor and write but now <u>recline</u> in my berth. And the eggs which we had for breakfast are not ordinary or common eggs, for three or four at mess had real chickens in them — fully fledged. But still we are being well fed in marble halls too.

Mediterranean Expedition Force

Dearest

We're fairly on the sea now, rolling, rising and falling all the day long and now it's night, black with no light in sight. All day we have not seen more than 2 or 3 steamers, and half a score of fishing vessels early this morning, and no sight of land since we started. It was dark when we quietly slipped out, a few vessels nearby signalled good luck, good voyage or bon voyage as well as some land signal stations. A few steamers gave us a farewell blast on their whistles and we were gone.

Last evening before starting we lay in the harbour all ready to start. It was a beautiful evening, the sun setting far out at sea gave a grand effect while the band played selections one of which was "Where my Caravan has Rested". Numerous seagulls hovered round the vessel, it was all very beautiful.

We are well fed and very comfortable and there's no fear of harm. We wear our lifebelts all day long and have them in our beds at night.

This morning 4 destroyers steamed by our sides, it was a sight to see the waves dashing right over them making it appear for a moment as though they were under the water.

Have not as yet been seasick, many have though. Chas had a slight attack this morning but nothing much. He now lies reading in his berth just above me.

This evening I had to sew a button on, and I've got a sock to darn when I feel in the mind. The spirit hasn't moved me yet, I don't think I shall care about darning, and then I've got some washing to do.

Mediterranean Expeditionary Force
Monday 2.8.15

Dearest

It's 7o/c August Bank Holiday and the sun has just sunk into the west with the rock of Gibraltar. It's been a perfect day — not a cloud, warm and bright and hot too. It is said we are reported to have been torpedoed, but that the vessel actually sunk was the "Hibernia". The submarine is supposed to have been waiting for us but missed us and so we are peacefully sailing in the Mediterranean. From Saturday afternoon to Monday (today) we saw nothing until we approached Gibraltar. Then we first of all saw a few sailing vessels, then an occasional steamer and finally land appeared, Spain on our left and Africa on our right, mountainous coasts both of them and rocky too, grand stretches of sand and beautiful blue water. The Mediterranean suits me well. The band of the Hampshire Regt has been playing some very sweet music.

Passing Gib we were cheered by the sailors on board some British Naval boat.

I suppose you've been across the water today and playing tennis too. Well how did you get on, were you in form? I'm looking forward to a letter from you telling me everything. Charlie remarks just now if he could do so he would call in yours and ask for some raspberries and cream. So far as I'm concerned I'd be thankful for a lump of bread and cheese. Oh! We do like food - bread and marmalade for tea and not a crumb left and I'm ready for tomorrow's breakfast now. Still I'm very well and don't think we shall take any harm.

Tuesday 3.8.15

I'm running away, oozing through the pores of my skin, all day this has been going on and now what's left of me I've laid down in my berth. Early you may think. Yes it is, but it's dark and there's not much to see, and we turn out at 5.30 in the mornings and I'm generally ready for bed. I expect it's somewhat about 10o/c at B'sea and you're in bed too. What have you been doing today? I expect I shall hear in your letter. I suppose Doll has returned today looking bright and brown. No photo of us passing though eh? Is it hot in B'sea? If so I guess you've not done much tennis.

We've been zig-zagging very much along the coast of Morocco and Tunis, saw two British battleships of some sort in the distance and one or two steamers, probably Spanish. The sea is a beautiful blue, the sky bespattered with white and fleecy clouds. During the afternoon the band played — one of the pieces was "Take a pair of sparkling eyes" and "Where my Caravan". This one seems to be a favourite of theirs and sure it is beautiful especially in these circumstances. Last night just as we were getting to sleep the alarm sounded and we had to scuttle out and get on deck, with our life belts of course for we always wear them and must never be without them. It's not necessary to put into words my thoughts and feelings concerning you for you know and understand without even as I know and understand you.

Wednesday 4.8.15

It's nearly 11pm, I just went out to have a look at the night. It's a grand night and we're slipping along at a very high speed, and strict orders are given that no light must be visible from outside the vessel. I should think a thousand or so men are lying out on the deck, many in bathing dress only, some have worn nothing else all day every now and again as I write I pause to mop up the perspiration, it is so hot.

We saw a lovely island today — high and rocky, quite a high mountain with houses dotted here and there upon its slopes. The sunshine has been unclouded — blazing hot and the sea has been perfect.

With fondest love,
Yours ever,
Tim

PS You might enclose a sheet of paper and envelope when you write please.

Thursday 5.8.15

We are ordered to be ready to land by 6pm today but reckon on arriving at our destination about 4am in the morning. We are told that the sun and lack of water will be even more serious matters to reckon with than the Turks. I should be very thankful if you could send me a few sweets or peppermint or acid drops or something of that sort, not many you know, just a few in a tin, these will be very acceptable.

Have enclosed you a copy of our newspaper, just out. To appreciate the news therein one needs to be on board for the voyage, so I have written one or two brief explanations on the back.

Dearest

We only got 5 minutes to get letters written and handed in. I fear the censor may have stopped one or all of my previous letters - it is stricter than I anticipated.

I am quite well and safe and in good spirits and sincerely hope you are too. Give my love to all. Am looking forward to a letter from you.

Fondest love,
Tim

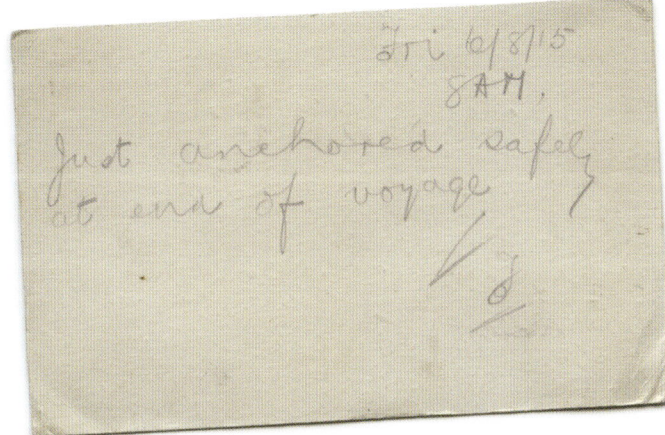

Sunday 14.8.15
Afternoon

Dearest

I've just been doing some washing in the sea. We go into the water and take our linen with us and there we stand and "scrub, scrub, scrub, at the washing tub" with no scrub brush though. You will gather from this I am safe and well, should like a cup of tea after dinner and come to tea with you as well, but am sorry I shall not be able. Time to write is short, we have had some fierce fighting and I fear many of our men are lost. But thank goodness some of us have a few hours respite and so it is I'm writing now.

As yet we have received no mail. I have just seen Noble Eagle [Olive's brother] and he is well.

Later

I wonder how it goes with you and what all the news is. What's the War news and what things have happened at home. You will let me know what you are doing and how you are getting on.

One gets accustomed to the roar of guns and the whistle of bullets. Even as I write Charlie lies by my side asleep, whilst overhead shells roar along to their destination. We're most of us a bit worn and weary, for hard work has been done and food as well as water is very difficult to obtain. This is a very barren land, no food, no water, no cultivated land and even shelter from the blazing sun is difficult to find.

Much of our tale must be reserved till we return, for although we have been here but a few days we have experienced much.

16.8.15
In the trench

I made the tea for breakfast whist Charlie made improvements to our "dugout". It's exciting work, for it is necessary to fall down flat every minute or so, or some shell might quench your thirst altogether. But I can assure you dearest I'm taking good care of myself and am not likely to get bowled over, and I am in good spirits too, if you care to come and have tea with us this evening we shall be very pleased to see you, only don't expect us to satisfy any hunger or thirst you might bring with you. I've seen Noble Eagle 4 or 5 times and have had several short chats with him. The last time I saw him he was well, his brigade was then in the reserve line and we are in the reserve trenches. He was very pleased to see me and of course I was equally pleased to see him.

No mail has yet reached us, and since we have landed here no mail has been collected from us and I don't know when we shall have an opportunity of dispatching letters, so don't worry if letters are few and far between. Depend upon it I shall write home to you as frequently as possible.

It's mighty hot, what we need most is Oxo cubes and hard plain chocolate. The water we get (it's scarce) needs boiling and something in it to drown the flavour.

Oh you should see our faces, we haven't washed or shaved for days.

Fondest love,
Tim

Sunday 22.8.15
Afternoon

Dearest

Hurrah! The first letters have just arrived. I was lying in my dugout, having a snooze when I heard the word go round. I was up in a nick and sure enough there was one from you and a PC from Blundell. You cannot imagine how we have looked for the mail, and now we've got it I cannot tell you how delighted we are, and no one more than myself. Do not curtail the length of your letters please, do not get the idea in your head that they are too long. Just write as you feel disposed and send it along according to your own sweet will and you will be doing much for which I shall be thankful.

I've just had an after dinner cup of tea — my word it was delicious. We speculated before making it as to whether we could spare the water. Finally we each put a few spoonfuls in the mess tin and boiled it after spending about an hour in search of a match with which to light the fire. Then we found a small piece of cloth in which we tied up some dry tea which we placed in the boiling water and made as I've said a delicious cup of tea — minus the cup though, and no milk, probably at the same time as Auntie was serving tea at home. You might tell her to have a big cup of tea ready for me when I return please and also a larder full of food. My mate says — 2 big cups please. Both milk and sugar, but we can drink it if necessary without either.

Really you would laugh to see us. Like foxes we live in the ground. We come up and walk about, light fires and cook food and make tea regardless of the occasional shot or shell passing by or over, until perhaps a heavy shell fire is directed upon us and like rabbits we all disappear.

It's Sunday afternoon — 2 Sundays have passed since we came to the trenches. It's no different to any other day. We do not know one day from another. Am sorry to say Charlie is rather unwell just now and is down at the base for 2 or 3 days. Nothing serious, a bit knocked up like many others. A day or so's rest and he will be fit again.

I've seen Noble Eagle again and he took me to see Clarence Kirby. Us three are so far as we know the only B'sea boys in this part. Noble gave me a packet of Woodbines, a box of matches and a drink of cold tea with a slight flavour of rum. We sat down on a rock and had a chat. Needless to say we talked of those we had left behind.

A mouth-organ eh? What are you going to do with that? Don't spoil the shape of your lips, will you. How am I you ask. Well just now I'm very well indeed, my cup of tea has made me feel very comfy and then there's your letter, what more can one wish for? Then the sea on my left is a beautiful blue and the sun is warm and bright. Why, if you were here we could be alright sitting on the hillside by the sea.

Please don't rack your brain to write that I may read and be madly happy, just write according to your inclination - happy or sad - and I shall be glad.

Am very sorry you were downhearted on August bank holiday, but am not surprised. Stick it Poll, the time to smile is sure to come, sure as the sun in the morning. But why weep? There like a passing shower it's soon gone and you smile again. You must get your mother to keep you busy, just tell her what I say - you must not be left alone to think. You must do some of the housework, or play your mouth organ, so cheer up Poll dear, we're giving the Turks snuff out here and shall be home again some day.

Monday 23.8.15

Little Wife

We've just dug our dwelling place a bit deeper for greater safety and soon we must see about dinner. We are in hopes of getting a drop of water with which to make a sort of stew: ¾ pint of water with a desert spoonful of dried vegetable and ½ a small tin of corned beef. Soak them and then if you can get a match to light a fire, boil it, and there it is, a tasty dinner for two. Oh! Put a cube of Oxo in if you've got one, come if you like, you can go halves with me. We've almost forgotten the taste of bread, biscuits are our bread.

24.8.15
Afternoon

This is where the body bands you made me come in useful - it's hot by day and cold at night. They keep warm where warmth is needed most for the sake of health. One thing I'm going to do when I return — that is kill all the flies that come near me, they fairly have their fling here, just because we've go no flypapers I suppose.

Now I'll finish up this letter ready to post the first opportunity I get, it's quite a job to get letters off from our dugout.

Take care of yourself dear for my sake, and remember I'm doing the same for you. Don't forget I'm always looking for the postman.

Fondest Love,
Your Man

Sunday 29.8.15

LW

We do live! 6 packets of Woodbines and a whole box of matches, and if I get half a chance I'm going to have a bathe this afternoon. Then tonight we go into trenches again. It doesn't matter much where you are, shells find their way to you. It's quite exciting, ducking and dodging and lying flat down — that is if you're not hit. Some cruel work was done not many yards from us this morning and something like 40 mules and horses are waiting burial, and 2 or 3 Indians are already in their graves and others in the hospital.

But still we keep very cheerful and you will be glad to know I am keeping very well and in excellent spirits.

The Indians are very interesting fellows — hardworking. They bring us food, water and ammunition. They are really good-looking chaps, refined looking, more so than I had imagined. Some have got beards which are of a most strange colour, bronze or a sort of golden sunset shade. Their work is to lead the mules by winding paths over steep hills to the trenches with provisions and ammunition.

This morning an enemy aeroplane passed over and dropped bombs on the hospital, and whilst out digging some trenches I saw some shrapnel explode over some Red X vans just returning with wounded to the hospital.

It seems incredible that in England today all is peaceful, folks dressed up in their best are taking quiet walks along shady lanes, perchance talking about the war from a more comprehensive point of view than ours and yet quite ignorant of what it is like even in so small an area as this.

I believe I told you Charlie was queer. Well he is better now and has been given a less arduous job with the battalion orderly sergeant, and so I don't see quite so much of him now.

Tuesday 31.8.15

Field Telegraph operating has kept me fairly busy the last two days and nights but it's interesting work and I'm quite comfortable in my little trench. Last night a large body of men moved slowly and silently past within a few feet of my place and among them was Noble Eagle. I had a couple of words with him and they were gone — swallowed up in the darkness. He was quite well.

Next Monday and my first year of service will be completed. You will recall on Sunday the happenings of a year ago and how on Monday by the first train I took with me a few necessaries.

I'm looking forward to another letter from you, have only had one as yet. It's a bit of a job to get the mail to us for we are in a very out of the way quarter and the limited means by which supplies reach us admits of little facility of the passage of mails.

Saturday 4.9.15

I could scarcely believe it when I heard someone say that tomorrow was Sunday. 4 days and no word written or posted to you. I believe that's the longest lapse I've made since landing here. What have I been doing? Goodness knows I've not been sleeping and even now my minutes are short. But few as they are they shall be devoted to you for it so pleases me.

The Chaplain has just passed with a word of cheer. He conducted a service on the seashore last Sunday. We missed many familiar faces and if tomorrow we have another service there will be others we shall miss.

But still we keep smiling. We make our tea, we cook our food and you would be surprised to see us really enjoying ourselves, and when I have a few quiet moments my thoughts come to you and I am really happy.

I do not know when I shall be able to send this letter, so far as I can learn no post has left us for a week, we're a bit isolated just now, but sure enough it shall be sent off first chance. It's just a year ago to the hour since I met you and Olive coming up from the Hard and told you of my intention to enlist. I can recall every detail, and no doubt you, even as I write, are doing the same. Ah! You wouldn't know me today — I've just had a wash and shave and cleaned my teeth and had a hair cut too. My first wash and shave for a week, and I feel beautiful and clean. But it's a job to wash linen in the sea. They won't come clean really I think I shall have to send my washing, mangling and ironing home to you for I've heard tell of your washing and ironing. We get good drying here, though. And darning I'm not much of a hand at

IN THE DARDANELLES

Brightlingsea Lance-Corporal's Experiences

The following interesting letter has been received by Mr. A. Portass from Lance-Corpl. T. C. Foster, who is serving with the 5th Batt. Suffolk Regt. at the Dardanelles.

"Did I tell you of our 'landing' how we came in a liner to a small island, and after waiting a day or two were transferred to a small steamer in which we cautiously zig-zagged to another place within the sound and sight of Naval guns and how for a night, a day, and another night we lay with numbers of other small transports, and then on the morning of the second day the order came to steam up, and how in an hour or two we came to those places where only two or three days before the first men had landed with many losses for a land mine had killed a number. We waited all the afternoon while several warships hurled shell after shell at the enemy entrenched in the hills sloping up from the sea. Our time came at last and we were again transferred to a very small steamer which by careful manœuvring landed us on some rocks where as yet no proper landing was made, and then we stumbled along in the dark—for it was between 8 and 9 p.m.—over some very rough land—chiefly sand and rock covered with prickly bushes for half a mile and were then halted and given orders preparatory to moving up into the reserve line whither we struggled and there we spent the night. Water was our most urgent need, it came to us from the base by mules led by Indians, it was "distilled" sea water from the ships. It was precious, if unpleasant, for if we got a pint and a half a day or two pints that was all, but we afterwards found a hole near by into which water found its way, it was thick with a greyish mud, but our Major assured me it was good because there were frogs in it, and so we boiled it and put tea in to drown the flavour.

Now we have moved along and are working with the Australians, New Zealanders and Gurkhas and there are other Indians too and water is good, although difficult to get. Wells have been sunk and excellent water found, but it is too risky to go by day, so we go after dark and wait an hour or so for our turn at the hose, and then we struggle back over very rough ground, our only danger being bullets which whistle over and around. Only the night before last a couple of Turks were captured near the well with poison and now it is well guarded. Last Sunday morning as daylight was dawning a small party of men passed on their way to the well for water when one of them was shot. All the time I was within a few yards. Many have lost their lives in fetching water. One place I shall never forget over a piece of high ground terminating in a high cliff. A canvas tank had been placed on the seashore; by day or night naval boats would fill it with water and here the men who were entrenched near by would go for water. The Turks soon made a mark of it, and it was well nigh an impossibility to get over the high ground alive, and so we took to going down to the water's edge through the gully on our side of the high ground, and from there wade or swim round the cliff to the tank. Even then a few yards was open to fire, and one morning seven men were killed at this spot. I've seen three lie dead and one killed, and have narrowly escaped the same fate myself. I could write pages of such instances, water supplies are all marked and although every care is taken lives are frequently lost. One day whilst fetching water we were shrapnelled, the fellows in front and behind me were hit, but it was untouched with the exception of a little dust.

No doubt you read of a 'new landing' in Gallipoli and thought of us. No details are given, but by studying the map you may perhaps locate the place. We have 24 hours in the trenches and 24 hours out, that is 24 hours in fighting order and then 24 hours out perhaps, for the probability is that the greater part, if not the whole of the 24 hours out are spent in "fatigue" work which probably takes us back to the trenches as diggers in parties of 50 in 3 reliefs, and the digging may be anywhere in the firing trench or communication trench. Wherever we may be or whatever the duty, if it be in the trenches it's not pleasant.

If in fighting order the probability is, it will be 7 o'clock in the evening when we move, carrying with us rifle, 200 rounds of ammunition, water bottle, respirators, and great coat. The entrance to the trench is hidden from the enemy, and can be made as safely by day light as can be made as safe by daylight as there are fewer stray bullets about them. We enter in single file and as we advance the windings become more pronounced, until at last it's a perpetual zigzag. Now and again a pair of heels project from the bottom of the trench or perhaps the whole foot is exposed, the constant shuffling along of the living is disturbing the "rest" of the dead. Some say "That's a Turk," they know by the boots, or an Australian, by the little bit of cloth visible. As we advance the atmosphere changes, and we give thanks for the cigarettes and tobacco which are served out, for the stench becomes intolerable, and a glance through the periscope will show why. For days and perhaps weeks a large number of dead must be unburned, even beneath the earth thrown up on the parapet the dead lie but lightly covered and often partly exposed.

The remainder of Lance-Corporal Foster's letter which contains some very exciting incidents will be published in our next issue.

Brightlingsea Times 1915

and sewing on buttons is a nuisance. My position is not very comfy — half lying, half sitting, but if I get into any other position a sniper keeps popping away at me and I've just picked up a bullet which struck the ground by my side just as I started to write. Really I think I'll get somewhere else...

After climbing about and scratching myself I've found a safer and more comfortable place but had I known how the flies swarmed here I would have brought a net to get inside of. Why, they get into your mouth after your food and if you linger long over a meal they will consume it for you!

We've got all sorts of Indians here including Gurkhas, they are strong fellows, very clean and courteous and civilised. They are all very interesting. Their dress, manners and customs would fill pages more than time permits me to write.

I cannot tell you how eagerly we are looking for a mail. We've only had 2 since landing and the only letters I've received have then been 1 from you written or posted 2 or 3 days after our departure - 2 from home and a PC from Blundell. Of course they'll come along in time but I hope it well be soon.

It's now evening, and cooler. The flies are disappearing and really it's quite pleasant. How I should love to be with you, what a walk we could have. I often look forward to the time when we shall meet again, surely nothing shall part us again. But now I live again the time. We have had such happy times. I am thankful, for the memory of it often makes me glad in this dreary land.

Fondest love Darling,
Your Man

Monday morning 6.9.15

So accustomed have I become to digging myself in wherever I go, that when I come home again my first job will be to dig a place of safety for myself in the garden and cover it over with bushes for shad. Fancy being able to walk about with no one shooting at you. Why, I shall be dodging shells all the way up the street. But to be able to get anything you want for money, why it will really upset our equilibrium. Ah! I'm saving money here, haven't drawn or spent a halfpenny since landing.

How's Dorothy Mary? And how's her nurse? Do you remember that morning when after breakfast we went of a walk round past the Weslyan school. This is just such a morning. What say for another stroll around?

All my love,
Tim

8.9.15

Little Wife

It's a job to keep count of the days, whether is Tues or Wed I know not. As I write an aeroplane whirs overhead and I can hear the sound of bursting shells round it, but really I cannot get out of my dugout to look at it. Last night we were in the trenches. Trench bombs and bullets were the only disturbers of the peace. Really it seems most strange, on my right I could look along the trench and see a most glorious sunset, far over the sea beyond a high mountain. I cannot describe its beauty, and then as the darkness deepened the lights of a hospital ship shone out — a row of green lights with a cross of red lights in the centre.

It was a fine sight, it brought to mind pleasant evening walks in country lanes and across fields and meadows where poppies splash corn with a glorious red and grass is a beautiful green. For a moment only one could reflect and recall happy days and breathe this atmosphere of beauty and peace. In this moment the spirit of one I love seemed to mingle with mine, and in an ecstasy we breathed as one in a heaven of our own. For a moment only, yet it was worth all the weary struggle of a hot day. Loud explosions, bright and blinding flashes of light and the unceasing crack of rifle fire with the whistle and whir and then the plonk of bullets as they struck the sandbags or the earth just above one's head, recalled my wandering fancy almost instantly to the firm realities surrounding us. But my spirit was refreshed and strengthened and some hours later when I was able to snatch a few minutes' sleep it was with you I rested or wandered in a dream. In this way you are playing your part through me, would that I were less unworthy, for it's here in the fight that one becomes aware of one's weaknesses. It's easy enough to be strong and brave at home. Come out here and try your strength and exercise your courage, put it to the test, and you will discover its value.

I fear my letters to you are poorly written and not exactly ideal love letters, but I know you will make allowance for all this. They are written for you, although I do not mind who else may read them so long as you make allowance for all my shortcomings.

And now I must have a look at my rice or it will burn and my dinner will be spoilt.

Good bye Little Wife,
All my Love

8.9.15

Darling

You are a pet – such a long and lovely letter. I was quite disappointed when I came to the end that it was not 4 times as long. Why, I feel as happy and light hearted as I have never been since I left you! Two mail bags were brought along on a stretcher and soon it was in my hand. And what a time the flies had, they well nigh devoured my rice and as I read I spooned away without looking at what I was eating. But that didn't matter, Poll I nearly danced for joy.

Little wife – you're a brick, and if I have to go without sugar in my tea you shall have the black shiny frames and the black and pink curtains and of course we shall have a smoke together whenever it pleases you and we shall be very comfy. Why, I can almost fancy it now, the room as you have planned it – the fire – no other light – and you, darling – my little wife – so close and cosy. And then this bedroom, I leave it all to you to plan. All I ask is for a place by your side. I don't even ask for a pillow for my head so long as you find me a place on yours... What a lucky chap I am. Why, it would have to be some great reward for me to enter such a weary struggle as this again and yet I could accept it 10 times over for you! So much do I love you. Your letter has lifted me into a heaven, into a state of mad happiness.

But you've enclosed no paper or envelope.

I've been puzzling my head as to where I shall be able to get hold of a better job when I return but can't think of anything yet. It's difficult to foresee the state of affairs at home after the war, am glad you realise we shall not be multi-millionaires. You must go in for economical cookery. Don't forget I'm not much of a meat eater. Dairy produce – fruit and vegetables is my line. Why, a piece of bread and cheese would come to me now as the height of luxury and would by a good dinner specially if I might indulge in an onion! I'm glad in a way we can't spend money here – every penny counts, don't forget that Poll.

Now my love, let me proceed to reply to your letter. Certainly, my sister in Australia [Millie] would be delighted with one of your photos. Poor little pal, you must take care and not knock yourself about while I'm away. I hope you are better by now.

You dreaming too much? Have I not told you to dream on, of course I have. Dream on and I'll dream with you, and together we'll realise these dreams. And don't think I shall "sigh" when you put your thoughts and fancies down on paper for me to read. Put them down Poll and I will do the same and thus we shall come nearer to each other and become entwined together, for if we freely mingle our thought and fancies, our actions will become harmonious.

You tell me the soldiers sing "Lead Kindly Lights". I can quite understand its effect on you. It sounds like the lament of someone lost. I have heard it sung out here, softly - because of the enemy - and just as darkness falls. "The night is dark and I am far from home" is sung from the heart. Truly it sounds like a lament for the night is dark in more ways than one here.

Am very pleased to hear you are making yourself so useful, helping others. You will be able to do much when we settle down and I know you'll be twice as eager when that time comes. Bandages for the Dardanelles wounded eh? Goodness knows they want them. Wait till we get comfy in that black and pink room, with you nestled up closely by me. Then I'll tell you of my love for you.

Yes dear, I know there are times when you are weary. There are times when I am weary too but I stick it and smile because I know you are doing the same and the thought that you are doing so helps me too.

It's very good of Marjorie Ainger to talk so - quite right! Take care of yourself for me - doing the same for you and I am also glad you told your mother you were very happy. Never give her cause to regret consenting to our engagement. Give her all the cause you can to be glad she did so. Am very glad Blundell wrote you, of course being a friend of mine he would not say anything against me. But really he has not got a girl - his love affairs are somewhat tangled. Although he is really engaged the engagement is suspended and he has fallen in love with someone else who is already engaged to another. So his love affairs are somewhat mixed.

Evening 10.9.15

I didn't write yesterday. I didn't feel like it. I like to write best when it is cool and quiet and those moments are rare. This however is one of them. The sun has disappeared below the hill which protects us from the west, and the bombardment which has been going on and off all day has ceased, and save for a bullet or two and an occasional bomb all is quiet. There's been some shells over our heads today, some roar along like railway trains, others travel with a rushing sound. They can be heard coming, passing over, and going on to their destination, and then comes the sound of the explosion. But those which drop and explode on us are scarcely heard before the explosion takes place then we have to be quick, and this morning we were continually falling flat into our dugout as shrapnel flew all round. It was quite exciting, specially when pieces of shell and shrapnel bullets went hustling through the branches shading our dugout from the sun, knocking the earth down upon us and, hang it, making the jam on my biscuit all grit and dirt!

11.9.15

Another evening, cooler and calm after an uneventful day. Last night there was a hail of bullets just before I went to sleep but my sleep was sweet and we were up early and what a treat! Bread was served out for the day and in a very short time I had fried my bacon and some of the bread and, my stars, you would have felt hungry had you seen it! It was a tasty dish and I enjoyed it. Well, little wife, how goes it? There was a small mail in today but nothing much. Now I must be off to fetch some water.

12.9.15

My stars what a mail - 5 letters and a newspaper! 1 letter from you posted 20th August, 2 from Blundell, 1 from home and 1 from Bushey nr Watford. Hurrah! Dashed if Charlie hasn't brought me another letter from you. Noble Eagle landed at ... before coming on here and he could send a letter from that place, whereas we came here direct, landing nowhere until we reached our ultimate destination. So his wife received a letter before my wife. He did not see anything of our battalion until we were in the firing line.

Thanks very much for sending the sweets. No doubt they will turn up in a day or two. I believe the smaller a parcel is the quicker it gets through. The apples too will be very welcome. I'm looking forward to the parcel and hope it won't be long before it comes. I have heard talk of restrictions on parcels for this part but there were several small ones came through today for different fellows, and anything accepted by the PO should so far as I know come through alright.

It was rather sad to see the mail sorted out today (Sunday). They were sorted in heaps — all the killed in one, the missing in another and the wounded in another. There was quite a large heap for those who have been killed. Their names were called but not answered, we knew them all, our companions, by the side of whom we started out. Some of them we even saw fall. How little the writers thought that those to whom they wrote would never read again.

Have you not made a mistake? You say last year 21 August I first put on your finger that little blue and white ring. Oh yes! I believe you're right. I was thinking it was 7 August and not 7 September I enlisted.

Good night Little Wife.
Fondest Love,
Your Man

16.9.15

In a gully high up among the hills I sit and write. It is evening. Far away from this high position I can look up and see the sea and through a soft mist dimly descry some high islands looming up out of the sea. Not more than 200 yards away — just over the other side of the hill — are the Turks in their trenches. Through little apertures in the parapets of our trenches our men stand and watch for a Turk to show himself and then fire at him, and they do the same to us. In the rainy season which is due very shortly I expect this, as well as all the gullies which run all over the place, is filled with water rushing in a torrent to the sea. I believe the rain when it comes falls almost without ceasing for about 6 weeks and is followed by winter. The weather here has already turned cooler and some rain occasionally falls, making it very difficult to get a fire to work and make tea. Sure enough the summer is dying and my thoughts turn to fires and drawn curtains and quiet evenings round blazing hearths. Even now as the daylight fades I look forward to the time when I shall come home to you and to happy evenings and sweet sleeps. Now I must stop for it's almost too dark for writing and like you I'm not allowed a light to go to bed by, and it's just started to rain and if I don't get myself together beneath my waterproof sheet well — I shall get wet and have to keep wet I guess, and have a mighty uncomfortable sleep. Good night Little one.

17.9.15
6.50pm

Now would you like me to write you again this evening, as the "Golden Sun sinks in the West" or would you rather I didn't. I'll presume the former to be the case and just write as it pleases me. But first of all I really must get nearer to the entrance of my dugout that I may be in view of the sun as it sinks into the west far over the sea. We have watched it together and seen the golden path stretching out over the water, and now as I glance up and see that same path it seems a path to you, for sure that's where you are, beyond the bright yet delicately coloured haze which deepens in colour as the sun lowers.

I'm always singing to you the same old love song, does it not weary you?

Now how goes it? Are you weary, sad or downhearted, or are you in high spirits? I'm quite happy. I generally am when engaged in writing to you. I only wish you could be here and with me to see this glorious sunset. I fear the shells roaring just over my head might frighten you but you would soon become accustomed to them. I wonder what you are doing and what you have done and what you are going to do. Tell me everything dearest. I'm looking and longing for a mail to come and bring me word from you. If you care to send a B'sea News occasionally I should be very glad to read it no matter if it is a month or two out of date.

The sun has just gone leaving a most beautiful sky behind and, if I'm not mistaken, the hospital ship is just sailing from her moorings. I expect she's got a load and will be replaced by another ship tonight.

Shall I tell you how I sleep when I get a night down? Well, I take off my pullies and boots and sometimes even my socks, also my tunic, and put on my cardigan. Then I don my sleeping helmet, wrap my blanket round me, tuck myself in top and bottom and bed down on the ground and the last thing I hear at night and the first thing in the morning is the slash of bullets from the rifles and Maxims as they fly over the valley in which we lie. Generally my sleep is so sweet and peaceful, no bullets or bombs disturb my rest. Yes, your prayer for God to bless me is answered.

Have I told you of my helmet? How it was ripped open by shrapnel a day or so ago? Lucky job I hadn't got it on, I'd only just taken it off and placed it on the ground by my side. Ah, it's made the dust fly! My ration bag was ripped open and my biscuits scattered, and a tin of bully flying through the air, bruised me slightly on the knee.

20.9.15

A very nice little letter, and it's postmarked 7.30pm 26th August. A funny thing, it only arrived last evening just as darkness was falling and I had to make haste to read it or I should have had to wait until this morning. Yes - it's a funny thing because I had one from you dated 31st August 2 or 3 days ago. And wasn't I excited when I heard of a mail yesterday, and weren't we all excited! It was about 2.45 on Sunday afternoon, a long string of us were on the way to get water in petrol cans. We had just passed through a trench which gives us cover in passing over a high ridge. We are not allowed to go along here any more than 1 at a time and then we all wait in a place of safety. Well, this is where we were when a couple of stretcher bearers returning from the base brought news of 6 mail bags for us. We went on our way and returned, the only talk or thought being the mail. Was it only a rumour? Would the parcels come? And when we reached our bivouac again we prayed the Turks wouldn't start playing the fool until we got our letters, and we quietly counted on parcels too. Time passed, shortly after tea time - about 6o/c — the Turks livened up and I thought they were going to be a trouble and thereby mess the mail up, but they quieted down again after wounding one or two, and your letter came to me as I have told you. Now that's a long yarn isn't it? And now I will proceed to answer your letters from where I sit, within 100 yards of the Turks - may they have the goodness to be quiet for a couple of hours.

You think it strange you writing me every minute? Not at all, I do the same to you. Do you think it strange? Yes, quite true, I am downhearted and dumpy sometimes. No cure you say? Oh yes! Just send me a very small tin of potted meat, only a small one, for I fear a large one might take a long time to get through, and besides when I open it it would be better to finish it for I've no larder in which to keep it once it's opened. Yes! I think that will do nicely. Pack it well. If I'm ever downhearted again, well, you can send again. Of course if you haven't got any potted meat send chocolate, and if you haven't got chocolate, send toffee. Anything in the food line. A cold sausage on a fork with a nice bit of bread and some mustard too. Oh, for a lovely piece of bread! Ah, and some butter too! Haven't seen a piece of butter since old England faded away into the night.

Oh, I must tell you how we live and how I cook! Bacon: ½ an egg once a month, bully beef, tea, jam, bacon grease, rice, sometimes prunes or plums (dried), sometimes dried vegetables/fruit biscuits. My word, you've seen them I believe. Well, live on them! Spread jam on them, save your bacon grease and spread it on them and you'll live something like we do. I'm a cook - a cook for my companion and myself. Bacon I'm quite an accomplished hand at but pancakes are my speciality. I can cook pancakes, nobody disputes this. In fact, others come to me for advice or tips. Beautiful and brown and frizzling hot, with a little sugar they make a dish far and away superior to anything out here. They fairly take you back to old England. Yes, they serve us out with 1lb of flour once a week. I mix it with some water to a nice thin paste, then I take a little bacon grease which we save from our breakfasts, and put it in the lid of my mess tin. Then pour some of the paste in and there you are. Some of the fellows make meat puddings, others make roly polies.

We get fresh meat once or twice a week, I generally get an ounce or two, and have beefsteak and onions. That's another luxurious meal. All this mind you is done on very limited means - a mess tin acting as saucepan, frying pan, plate, cup and table as well. And perhaps your water for the day is limited to a pint or a pint and ½. Oh, we don't get drunk on it! For, mind you, we probably have to wash and shave out of the mess tin too. Yes, it's our bowl and bath as well as shaving cup.

But I'm wandering. I started out to reply to your letter and had just reached the part when you said something about the dumps. Now I'll wander back again. Have you still got the same two soldiers - Sgt Alcock and the other chap? Give my kind regards to them if still with you. My head? Well a piece of wood cracked it on board ship, but it soon got better, and I haven't had a headache since leaving England.

Thursday afternoon on Underwood's Hard eh? Yes, I know and can almost feel the lovely breeze fanning your cheeks, would that I were the breeze! Ah, I should have loved to be there and carried the lunch for your mother! And the bathe too! I would even have suffered a mixed bathe to have had just one dip in the water. You must have looked very nice in white and red. Did Doll take your photo?

This concludes my reply to your letter posted 27th August. Now for your letter received a day or so ago and dated 31st August.

No, I haven't got your parcel yet but live in hopes. One of our chaps got a fair sized parcel yesterday posted on 9th August. Small letterpost parcels seem to be getting through alright, but I should think the larger parcels will come along eventually. Parcels should be small and well packed for they get a lot of knocking about.

"A very cold breezy night, Mother and I are settling by the fire having supper". My stars! Shouldn't I simply have loved to be there too, having supper by the fire! It's getting a bit cold here at night and if we lie down to sleep we have to wrap ourselves up well.

Yes, we were in the thick of it within 12 days of leaving Watford and have been in the firing line nearly all the time since. An advance which took place 2 or 3 days after landing cost us some lives — we lost our officers and a lot of men. It was over Thursday. I was at the Battalion Headquarters on telegraph work and had a shave in the morning. The Colonel and Major asked me if I would shave them too, for neither had brought their shaving tackle with them. Willingly I did so and pleased them. They were jolly good fellows both of them. That evening we advanced on the Turks and I've never seen them since. In the papers just received from home and dated round about 27th August I see they are both in the casualty list as wounded or missing. I also see a number of our officers among the killed and wounded. Shortly I expect the papers containing the casualties among our rank and file will come along. The 5th Suffolk will have a long list I'm sorry to say, almost every day we lose one or two, either killed or wounded, something like 30 to 40 of my companions from Hadleigh have gone one way or another.

That's right, keep happy and well for me and write to me as often as you like.

Now I will concede, the Turks have behaved themselves very well and I am thankful. Oh, if you don't send me a bit of pencil — indelible please — I soon shan't have anything to write with! And don't forget writing paper, and envelope, and an odd PC or two please for luck. You see some of your paper come back to you this time.

Fond Love Little Wife,
Your Man

PS Your 26 Aug letter is sweetly scented
Have enclosed a piece of my helmet which was ripped by shrapnel.

October 1915. One dreadful day I received a letter from Hadleigh from Tim's father telling me news from the War Office had been sent saying Tim was in hospital with typhoid fever. 3 more notices followed: seriously ill... dangerously ill... etc. These were dark days for Hadleigh and B'sea, Tim's father writing to try to cheer me up and I writing the same to Hadleigh.

We began to think we would never see our Tim again. Then at last we heard from him! He was then in hospital at Alexandria, and to our great joy later sent to England to a hospital in Edgebaston. 100 patients on the hospital ship coming over died. How lucky Tim was.

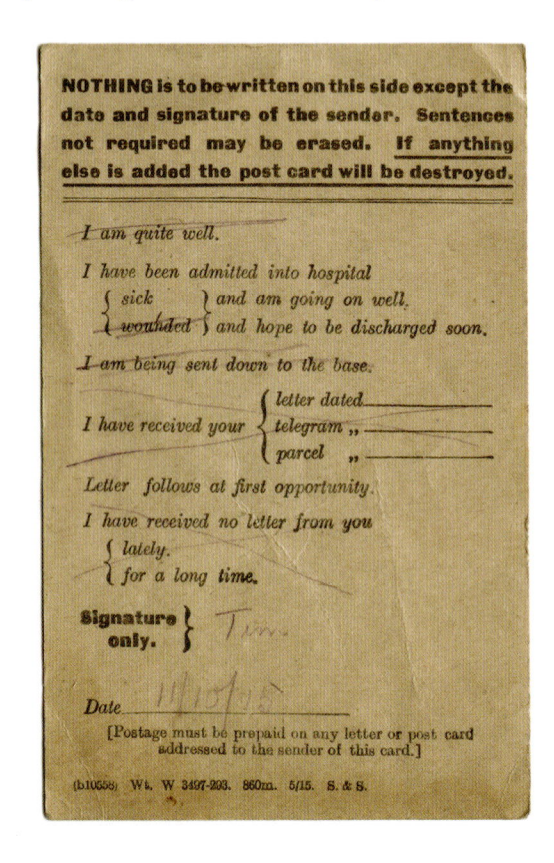

From Pollie's scrap book: Mustapha Detail Camp, Alexandria. Men were sent from the large General Hospitals here to recover from wounds and illnesses.

Ward 15
21st General Hospital
Alexandria
Egypt
31.10.15

My Love

However long is it since I wrote you? Great Scot, I can't remember! You must forgive me, I've got a good excuse. I've had typhoid, and for some time it was not possible for me to write. I'm only just pulling myself together sufficiently. For over a month I've been queer, and the arrival this morning of letters from you coming at a time when I'm just pulling round brings home to me the things I have left undone. You must forgive me dear — forgive me if this letter does not come up to your expectations, for I am not able to do much yet, even letter writing tires me. This is my 2nd day out of bed. Yesterday I was allowed up for ½ hr in the evening. Today I got up at 4pm. Oh, it's been a weary business lying in bed day and night for weeks, and I'm mighty thankful to get up if it is only for an hour or so! I really cannot remember the last time I did write you, whenever was it?

I will not attempt to answer your letters now, there are 2 or 3 of them, nice long letters, dear, for which I am very thankful. They cheered me up wonderfully and I am looking forward to others to come. I'll make up for lost time as I get better dear, but I must repeat how glad I was to get your letters this morning. Even though you may not have heard from me I do hope you've kept on writing me, for it does cheer me up so to hear from you, for my thoughts are ever with you and it's like a smile from your dear face to get a letter from you. As soon as I feel up to it I will answer your letters for they contain much for me to comment upon.

I expect it will be a week or two before I am released from this place, then I should not be surprised if I were sent to England for a few months convalescence, so I may be with you for Xmas yet. But this is not certain so I must not count on it.

Now Little Wife I must conclude and get to bed — fancy being out of bed! I laid in bed so long I scarcely realised I'm out — or I shall be sending my temperature up and I shall have to stay in my bed again.

Forgive this poor letter dearest, I'll try and do better next time. I trust you are well also all your people — give my Love to all.

Good night dear,
Your Man

7.11.15

I'm sure I never know what the day is but today something tells me it's Sunday and upon enquiring I find sure enough it is. My thoughts take me to B'sea. I can picture it, so peaceful. It's now 9am. The streets are deserted, save for an occasional pedestrian, perhaps sauntering up to the PO to read the war telegram. Victoria Place is quiet too, am I right? The leaves have turned and some are falling. The last tree but one from the "Duke" end of the green holding on longer than the others. It's July weather, bright warm sunshine all day long and not many yards from where I lie is the sea - deep and blue - although I cannot see it as I lie in bed. I shall see it when I get up at 11o/c. For I'm told this I may do - (thank goodness) Ah, then later on - after the services are over comes the parade! Now I hated that, doffing your hat every minute, yes I detested it. Can you remember me meeting you and O in Colne Rd one Sunday at this time and telling you of my intention to go to Colchester that very afternoon to make enquiries as to enlisting? Ah, of course you remember! I didn't like B'sea Sundays. With the exception of the evenings when dark, and you used to take the walk up the old Church Rd with O and her sister, and so did I, and thus we stole some sweet moments. Shall I ever forget them. How disappointed I was if I didn't see you, but it was not often we missed eh? Oh, I smile to myself at the manner in which we stole a few brief moments together in those days and feel a sense of great satisfaction at the liberty we now enjoy! We're lucky Poll - or rather I am. Now I've got several other letters to write and I will conclude.

With fondest love to
My Little Wife.
From Her Man

18.11.15

Dearest

Hurrah! I'm sailing for England tomorrow.
You must excuse this short note as I've much to do tonight.
Fondest love,
Tim

HMHS Aquitania
3.12.15

My Love

Is it really December? I was living in November until I glanced at the calendar. We have anchored for the night off Southampton and expect to disembark tomorrow (Sat) when 16 special trains will convey us to various hospitals and convalescent homes. Where I'm going I do not yet know, but will let you know early as possible. It's been a bit rough the last few days but thank goodness I'm feeling well and better for the voyage.

How goes it with you? I hope you are in the best of health and spirits. I'm looking forward to a few happy days in your company.

I'm impatient to get off this boat and set my feet on English soil again. Not that the English soil matters so much, it's "that friends the beloved of my bosom were there" to quote T Moore's "Vale of the Avoca" in that volume of "Irish Melodies" you have. Oh, I really must try and get to B'sea for a few hours! But you must come to Hadleigh too or I fear our time may be short.

Good Night Sweetheart

HMHS Aquitania
(from an old postcard)

Ward A7
1st Southern General Hospital
Edgebaston
Birmingham
5.12.15

I hope my love you did not mind paying for the PC or for this letter, for I can get no stamps and am not allowed to go out to get one. Well dear, I'm in college, what do you think of that? It's the City of Birmingham University that we're in. It's a large and fine place and I'm quite comfortable and getting on well. Send me a B'sea news will you? Just so I can read how it goes in the old place, and if you can I should like a glimpse of that letter of mine to Portass, published in the B'sea News. When you write please enclose writing paper and envelopes and next time you do any cooking make a cake for me dear. It's quite a treat to be in old England again and I should like to get as near to B'sea as I can that I may feel even more at home so don't forget me when you do any more cooking. I shall eat and fancy you sitting before me. Now I've given you quite a lot of things to do, I'm making myself a nuisance as soon as I'm back.

6.12.15

Last night (Sunday) the Bishop of Birmingham conducted a service in this hospital, it was the first time I've heard an organ since that Sunday outing when we went to church upon the occasion of my last visit to B'sea. It was glorious beautiful music.

After a while he got sick leave and came home to Hadleigh. He was told he would not have to go East again but would be stationed in England. That was good news. We enjoyed his leave at Hadleigh and B'sea. We were happy and full of plans and dreams. What tales he had to tell, which made everyone's hair stand on end, specially my mother. She could hardly bear to hear him, he got so excited and carried away, once he started to talk he couldn't stop. It all poured out, these tales of horror, the awful experiences he had, specially when ill with the enteric fever. He was given up as dead and, thrown out with a heap of dead to await burial… a miracle he was found in time and was then taken to a field hospital.

How different poor Fred was when he had another leave from France after being in hospital with shell shock. He just sat by the fire looking so sad & forlorn, unable to talk.

And then just as we were expecting Tim to be posted somewhere in England, he was re-called to the East again, there to remain until May 1919. Not to such dreadful hardship again: he was transferred to the Royal Engineers & appointed as station master to Haifa Station, Samakh, El Kubri & finally Jerusalem. Although wanting to get home, he enjoyed occasional leave which he spent wandering about the Holy Land. He got to know it so well, he acted as guide to many. He wrote wonderful descriptive letters of his experiences, and took many photos.

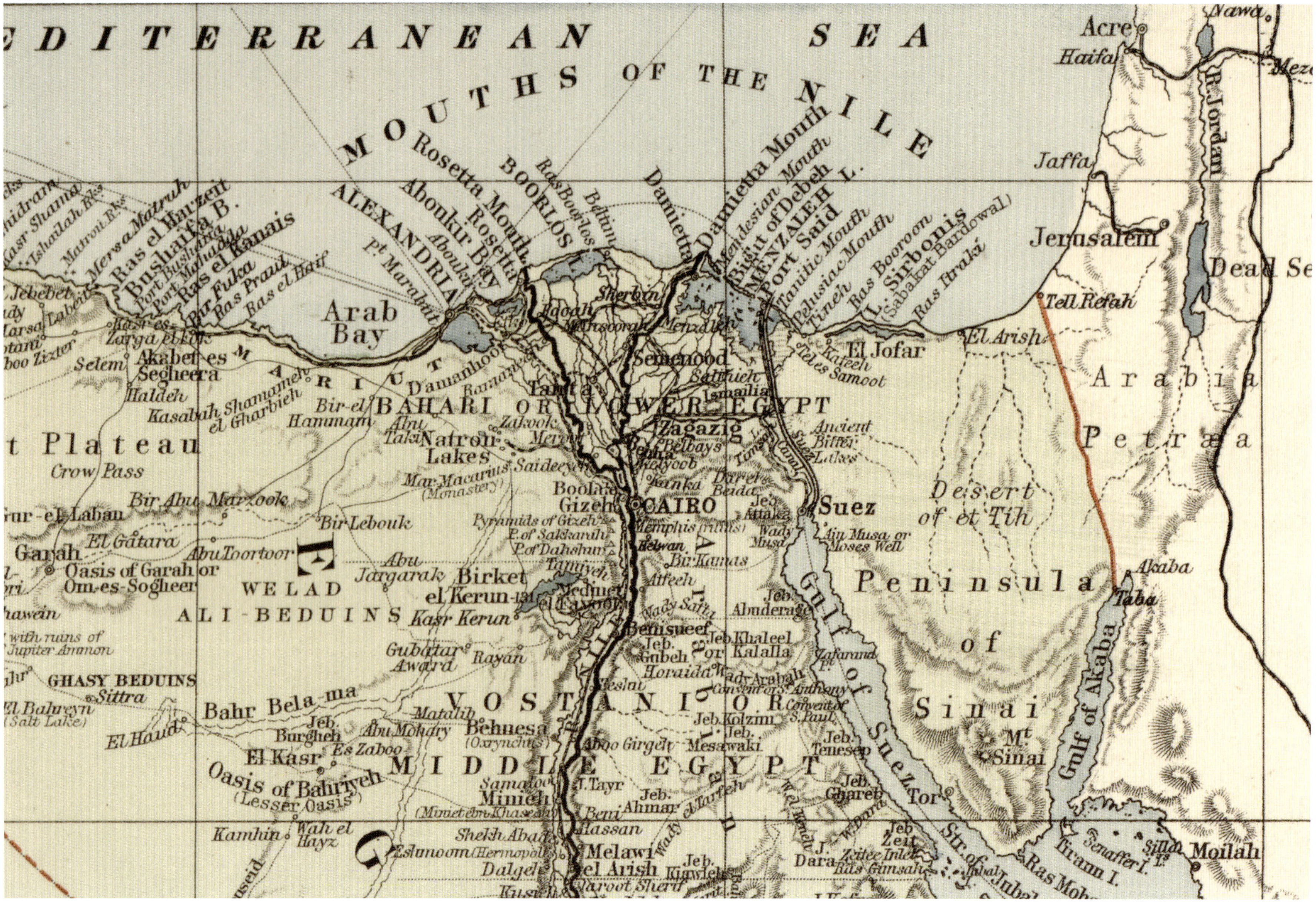

EDITERRANEAN SEA
MOUTHS OF THE NILE
Rosetta Mouth
Damietta Mouth
Mendesian Mouth
Tanitic Mouth
Pelusiac Mouth
Acre
Haifa
R. Jordan
Jaffa
Jerusalem
Dead Se
Tell Refah
El Arish
El Jofar
Arabia Petræa
ALEXANDRIA
Aboukir Bay
Rosetta
BOORLOS L.
Damietta
MENZALEH L.
Port Said
Bight of Debeh
L. Sirbonis (Sabakat Bardowal)
Ras Booroon
Ras Itraki
Arab Bay
Ras el Kanais
Ras el Harzeit
Bushaifa B.
Mersa Matruh
Ras Fuka
Ras Praid
Ras el Hair
Pt Marabut
Sherbin
Mansoorah
Menzaleh
Semenood
Tanta
Damanhoor
MARIUT
Akabet-es Segheera
Selem
Haldeh
Kasabah Shamameh
el Gharbieh
Bir-el Hammam
BAHARI OR LOWER EGYPT
Ismailia
Zagazig
Benha
Belbays
Tel es Samoot
Kateeh
Ancient Bitter Lakes
Natron Lakes
Mar Macarius (Monastery)
Crow Pass
Bir Abu Marzook
Gur-el-Laban
Garah
Oasis of Garah or Om-es-Sogheer
El Gatara
Abu Toortoor
Bir Lebouk
Gizeh
Boolak
CAIRO
Suez
Memphis (ruins)
Pyramids of Gizeh
Ain Musa or Moses Well
Desert of et Tih
Peninsula of Sinai
Akaba
Taba
Gulf of Akaba
Gulf of Suez
Str. of Jubal
Mt. Sinai
Tor
Moilah
Tiran I.
Sanaffer I.
WELAD ALI-BEDUINS
Abu Jargarak
Birket el Kerun
Kasr Kerun
Medinet el Fayoom
Bir Kamas
Atfeeh
Wady Sanu
Abuderage
Benisueef
Jeb. Khaleel or Kalalla
Jeb. Gubeh
Horaida
Wady Arabah
Zafarana Pt.
Convent of St. Anthony
Convent of S. Paul
Gubatar Award
Rayan
GHASY BEDUINS
Sittra
El Bahreyn (Salt Lake)
El Haud
Bahr Bela-ma
VOSTANI OR
MIDDLE EGYPT
Behnesa (Oxyrynchus)
Matalib
Abu Mohary
Jeb. Burgheh
Es Zaboo
El Kasr
Oasis of Bahriyeh (Lesser Oasis)
Aboo Girgeh
Jeb. Kolzim
Jeb. Mesawaki
Jeb. Tenesep
Jeb. Ghareb
Minieh
Samaloot
J. Tayr
Jeb. Ahmar
Beni Hassan
Kamhin
Wah el Hayz
Eshmoon (Hermopolis)
Sheikh Abadeh
Melawi el Arish
Dalgeh
Jeb. Kiawleh
Ras Gimsah
Ras Mohammed

1916: EL KUBRI LIGHT RAILWAY

On the 26th February 1916 Tim left England for Alexandria. The rest of the 54th Division had already arrived in Egypt straight from Gallipoli. Tim was on the *HMT Transylvania* by the 1st March, probably having embarked from Marseilles and arrived in Alexandria some time around the 12th March. From there he went to Cairo's Mena Camp within sight of the pyramids.

From the camp it was possible to take a tram to Cairo. Tim sent several picture postcards of Cairo to Pollie later on. It may have been here that he bought a camera and took the photographs that follow in this section. He developed the film using local water from the Sweet Water Canal, then made contact prints from the negatives using the sun. He would send the images to Pollie with a brief description on the back.

British forces had been sent to protect the Suez Canal, which had been attacked by Turkish troops in 1915 but who were unable to cross the canal or sustain themselves across the Sinai desert. In 1916 British defensive lines were put to the east of the canal in the Sinai desert to prevent any further attempt. Those evacuated from Gallipoli were joined by fresh divisions. All commands united to become the Egyptian Expeditionary Force (EEF) under Sir Archibald Murray.

Defence of the Suez Canal was divided into three sections (northern, central and southern). Each section had a bridgehead where a pontoon could be quickly attached allowing large numbers of troops to cross the canal from the railway on the west side, and a light railway to supply them in the trenches in the desert. Tim was sent to the Southern Section No. 1 at El Kubri, just north of Suez. He was in charge of the light railway (2'6" gauge) that ran from the bridgehead No. 2 on the canal's east side to the railhead in the desert, where goods were transferred to a 60cm gauge track with 2 Decauville engines on it. These engines supplied the two trenches "Salford and Oldham" that made up part of section 1 of the chain of defences along the length of the Suez Canal.

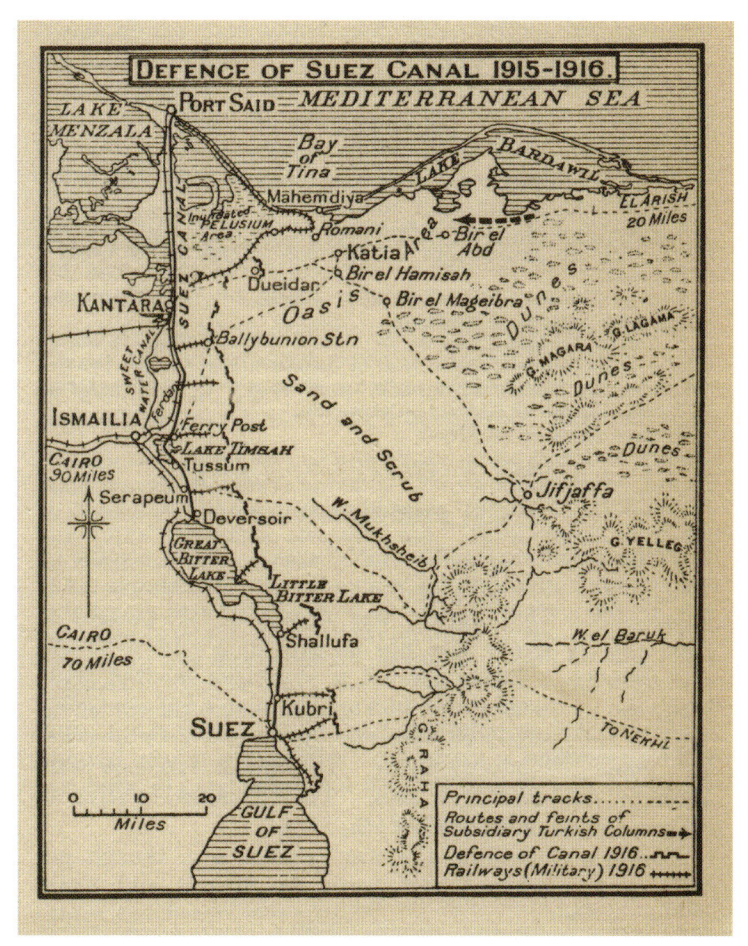

Left: London Atlas Map of the Nile Valley. London E. Stanford c 1910.

Map from *The Palestine Campaigns* by Lieut-General Sir Archibald P Wavell CMG MC. London Constable and Co Ltd. 3rd ed. 1941.

POST OFFICE TELEGRAPHS.

Aalton 6p

TO: Pollie Pannell
Moverons
Brightlingsea

Leaving for Egypt tonight letter follows

Tim

P.4

Thurs 2/3/16

Darling

There was no time to have my photo taken, I'm sorry dear, the order came so sudden & unexpected the notice was so short there was no time for anything.

3/3/16

Your letter No 1 posted on Mon was brought aboard this morning, I'm jolly glad we did not sail yesterday or I should not have got it until arrival

---

escorting us, the sea was the men were to say I was rather funny. I've got a ... I'm quite ...

easy matter your surroundings difficult. So may be it's sunny spring songs ...

bright pleasy ...

Hope Hope

Post Card.

Correspondence / Address

Embarked at 7 oc this morning on the "Transylvania" expect to sail 10 a.y. H.Y.

Miss P. Pannell
Moverons
Brightlingsea
Essex

H.M.T. Transylvania

Altho we embarked early
Wedy mng 1st we did not sail until the same day we passed

                                 1.3.16

   Embarked at 7o/c this morning on the
   "Transylvania". Expect to sail today.

   FL
   T

                                                    Thursday 2.3.16

   Darling

   There was no time to have my photo taken. I'm sorry dear, the order
   came so sudden and unexpected and the notice was so short there was no
   time for anything.

                                                    Wednesday 3.3.16

   Your letter No. 1 posted on Monday evening was brought aboard this
   morning. I'm jolly glad we did not sail yesterday or I should not have
   got it until arrival in ... It's very good of you to write me such a
   nice letter at a time when you were sad. I cannot tell you how much I
   appreciate it, thank you so much dear. I believe I've already told you
   I did not want to leave England. Poll my love don't think I want to get
   away from you when you are worried or in trouble. I had no choice in
   the matter or I should not have come.
   Last evening and night the sea was rough and many of the men were
   sick but am glad to say I was not. Today it's warm and sunny and the
   sea is calm. I've got a slight cold otherwise I'm quite well.

The Transylvania was taken over for service as a troopship in 1915. Sunk by torpedo 4th May 1917. Tim's telegram, postcard from the ship and letter heavily censored in blue pencil.

                                                    Sunday Morning 5.3.16

   It's an easy matter for me to picture your surroundings. Maybe it's
   sunny and the birds are singing spring songs, or maybe it's cold and
   dull. Here the sun shines warm and bright on a beautiful blue sea,
   fleecy summer clouds float above, far away on the horizon is a steamer
   with one funnel. All over the decks lie soldiers. I'm airing my
   clothing in the sun for I hope to have a bath today and have a change
   of linen.

                                                    Tuesday Morning 7.3.16

   We did not stop at ... and so I could not post you from there. For
   the past 2 days the weather has been terrific and today it fairly takes
   the cake. The wind shrieks through the rigging and the water sweeps the
   decks. Yesterday we were inoculated against cholera. Your body bands
   are a good safeguard against this.
   I'm happy and well and sincerely hope you are. When I lie down at
   night it's an easy matter to transport myself to you and live again
   those happy days and furnish again that little house.

                                                    Wednesday Afternoon 8.3.16

   If I post this now it will I am told be dropped at ... Am quite
   well though we have had some rough weather the last day or so and the
   sea has been very rough too. I'm looking forward to a letter from you
   at Cairo, where I expect we go. How is your mother now? I do hope she
   is better.

   Good night my Love

29.3.16

Dearest

Early yesterday morning I left our Regt and after travelling all day reached this place where I now write. It's hotter than ever here and the flies are a torment. 80 miles away lies the nearest town. Water is brought to us over the desert by camels, but although the sea is a long way away we are quite near salt water, and last night when we arrived I had a nice cool dip. It's very difficult to find a shady place and the sand with the bright sun blazing down on it is very trying to the eyes. When occasionally a sand storm comes up we are choked and blinded and everything is smothered with sand. In the cultural parts of this country flowers bloom and blossom abundantly, and most beautiful they are, but here in the desert they are rare. Only at an occasional oasis is there any growth whatever.

I enclose a very small flower which I plucked at such a place. There was one solitary palm tree, a few blades of grass and this was the only flower, one might wander for days and not find another.

"This bridge spans the Nile. We marched across it on arrival at Cairo."

30.3.16

We have tea at dinner time and dinner at tea time on account of the heat, and generally do most of our work very early in the morning or in the evening. We are not far behind the trenches but there is only spasmodic fighting.

The last evening or so I have managed to get a nice cool dip in the canal but I fear this will be out of reach in a day or so.

Sunday Afternoon
2.4.16

Another fresh place - a trip along the canal and here we are at El Kubri on the Railway.
Address me:-
2394 L/C TC Foster
(1/5 Suffolks)
EL KUBRI Light Railway Staff
54th Division
Egyptian Expdy Force.

I saw Chas a day or so ago but am a few miles away from him now and so do not see him. How are you my love? Well, I sincerely hope. One would scarcely realise here that today is Sunday, though I expect it's very Sunday-like at B'sea. If I were there we would go to Church and for a walk too if it was fine. It's fine enough here, the sun is blazing hot and the sand stretches far beyond the horizon all round.

I've just had a nice cooling cup of tea but it cannot compare with a cup in Victoria Place.

Life out here is very interesting though it has not the attraction of home. That's my destination, and yours too eh? We shall reach it some day and appreciate it all the more because of the waiting.

What have you been doing with yourself? Taking care of yourself I hope. Just write and tell me everything. I'm anxious to hear. This is such an out of the way place we hear no news bar that reaching us by post. Is the war still raging? Or is it likely to stop.

With Fondest Love
to my Little Wife,
from her Man

No. 2 Bridge at Kubri. Steel barge bridge is floated into position when needed to connect the two sides. It toc soldiers and 50 natives 5 to 12 minutes to float into position. "The eastern bridge head is 300 yards from my

Inset: Steel barge bridge from the west side of Suez Canal.

our Mess hut where we eat jam

"View of Kubri East taken at eventide from east bank of canal." Mess hut, and Engineer's workshop.

"Kubri East from east bank of canal. My shack is marked with a pinprick." The tents of the main camp in the distance.

HMS Euryalus steaming south past El Kubri Bridgehead. Tim's shed centre of picture between the tracks leading to the engineer's workshop on the right, and bridge head on the left.

"The Military authorities use native *Dhows* or *Gyassas* for distribution of materials such as timber, barbed wire etc along the Canal."

Bridgehead number 2 at Kubri East with "reef breaker" being towed by a tug.

Tim and his hut.

14.4.16

Dearest

    I do dislike washing - not myself but my clothes, and darning and sewing too. I've got all this to do but I've got no darning material neither have I any white cotton. If you should see any lying about you might send it on.

    Today we had a very heavy shower, the first since my arrival here, and for an hour or two. Yesterday we had a young sandstorm, sands obscured the sun which could be but dimly seen and it was a pale blue, it enveloped us like a thick fog for several hours. The wind was strong and carried the sand along, stinging and blinding it was and penetrating too, into our pockets and clothing, even into one's pocket book. We ate it and inhaled it and we swept it out of our ears and eyes, but I'm told it is as nothing to what we get sometimes.

    What's all this to do with a love letter you may say? Everything I would reply. It's the framework and canvas of the picture, but needs an artist, one blessed with the ability to convert feelings into form and convey those feelings to others. Now, I'm no artist. If I were I would pencil you a beautiful picture. Fortunately you perceive this, that I cannot write. Although I do at times launch out I generally get out of my depth and am lost, and lay down my pencil in despair. Then I get my two blankets laid out, wrap myself up in them and drift into the land of dreams. Day dreams are best, I often live over again the many happy days we had together. I recall our shopping expeditions in Ipswich and our evenings at home too. Really I must get enteric again. It's worth having, though I fear your folks may think my last convalescence should last us for some months yet.

    You never told me what became of that cushion cover. I'm looking forward to laying my head upon it in the days to come.

    How's Fred getting on and your mother too. Did Reg Fookes obtain exemption? The last letter I received from you was No. 7 posted 20th March and received 9th April. I surely hope No. 8 has not found its way to the bottom of the sea. I've not heard from Hadleigh or Ipswich for quite a long time. I hope they are all well.

    I could do with a weekend at Moverons now with some rabbiting and shooting but expect I must wait a while.

    Now it's bedtime -
    (Do you remember bedtime?)

Fondest Love,
Tim

"The Camel Corps washing on Sunday." The Sweet Water irrigation canal ran with water from the Nile, parallel and on the west side of the Suez Canal and provided drinking water. Tim used water from here to develop his photographs.

"Our shops' staff outside the shops. We have drivers from Canadian Pacific Railway, Buenos Aires Pacific, and men from various railways in England and a fitter from Burrell's Thetford."

"The smith's shop. Our handyman Sam from Cairo is Sudanese and is attached to us with others for platelaying. Our smith's shop is sort of self contained. We have our own fitters, smiths, carpenters, and wagon repairers."

Above: "Light petrol loco shed. Engineer's workshop with inspection pit just visible."

Right: Tim in petrol loco shed. Note "Station Master" armband.

"Avonside" petrol loco with a smaller gauge railway engine on a truck, a Decauville engine. The Avonside had a 4-cylinder Parsons engine of 60hp and 4-speed gearbox, top speed 15mph.

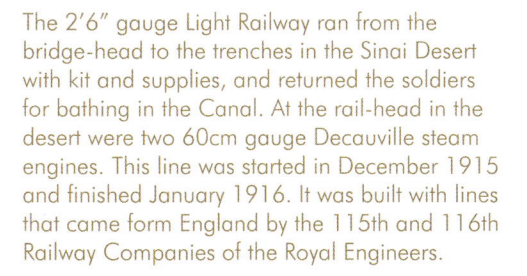

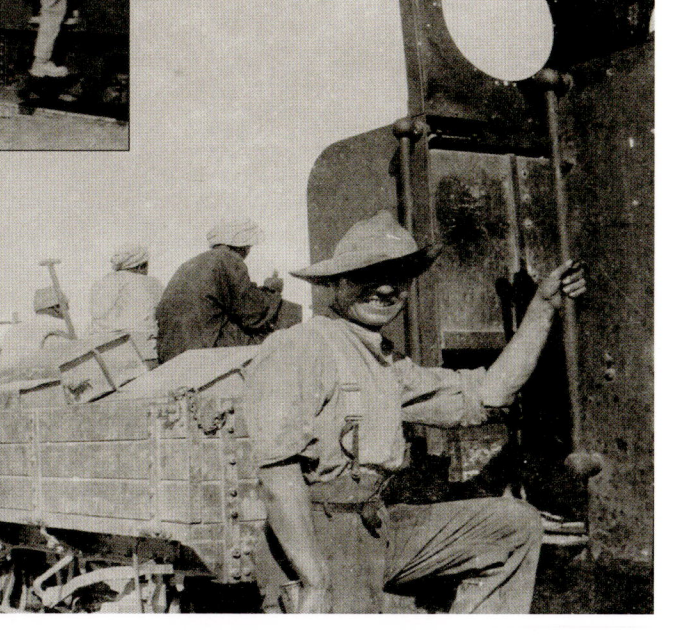

The 2'6" gauge Light Railway ran from the bridge-head to the trenches in the Sinai Desert with kit and supplies, and returned the soldiers for bathing in the Canal. At the rail-head in the desert were two 60cm gauge Decauville steam engines. This line was started in December 1915 and finished January 1916. It was built with lines that came form England by the 115th and 116th Railway Companies of the Royal Engineers.

"Loading one of our petrol locos for a trip up the canal. This one had done 3,000 miles in the desert."
Inset left: Floating crane unloading a steam roller.   Inset right: Two locos loaded on a lighter.

"Two "Decauville" 600mm gauge light steam engines at the depot in the desert. Used for distributing materials and stores to various posts ("Oldham" and "Selford") from the rail-head where our light railway terminates."

"Gurkhas leaving on the light railway to the front line trenches. The names of the trenches are Oldham and Selford". The Gurkhas were stationed between Kubri and Shallufa.

"Petrol loco just in from the desert front with party for bathing in Canal."
Inset: "No. 41 fresh on our line".

In the garden Victoria Place with Sergt Littlewood & Alcock

BRIGHTLINGSEA. G.E.R. STATION. No. 13.

The Australian and New Zealand Regiments and Royal Engineers were trained in B'sea for trench making and pontoon building. Trenches were made between Moverons and the Church. Pontoon bridges made and built over the ford.

All these men had to assemble on Lower Green (Victoria Place) for roll-call at 10pm. This was outside my bedroom window, and I was usually in bed at this time. Whilst waiting for the roll-call, the Welsh Regts would sing in wonderful chorus, maybe 200-300 men, song after song. I wished they would sing all night it was so beautiful. It would be heart breaking when they left for the front, singing as they marched to the station, everybody waving and cheering as they departed.

Several of these men wrote to me after leaving the town. I could easily have fallen in love with the handsome Scot if I hadn't already been in love with my Tim. Not one of them would I rather have had, so just as well I was faithful to the one and only.

At this time we had two sergeants billeted in our home, Littlewood and Alcock. A rather stodgy pair, they were with us for some time. My mother was very good to them. Alcock was rather an elderly man, in civilian life a schoolmaster. He used to annoy Doll at times, she hated him. One day at the dinner table she threw a tomato at him, caught him ploke in the eye! He was furious and never forgave her.

Two afternoons a week, my sisters and I went to the Manor House to make bandages and operation socks etc under the direction of Mrs Napier, wife of Admiral Napier in command of the Navy lads in B'sea.

And so the war continued, my time taken up with helping Mary, writing daily letters to Tim, and helping twice a week at the Naval hospital in the disused Railway Refreshment Rooms in Brightlingsea. I enjoyed making up & airing beds, keeping all prepared.

When at work in this hospital I used to find time to wander over the rooms upstairs, thinking what a nice flat they would make, writing to Tim suggesting how lovely it would be if the railway would let us have it when he came home. I even went so far as to plan placing furniture in the rooms. Little did I really think it would one day be our home.

Wednesday 6.7.1916

Dearest

No. 36 dated 21.6 postmarked Hadleigh reached me this morning. I am very glad you are visiting the old place. You do not tell me how you found your way there, and I should like to know. Tell me every detail please. You travelled on Tuesday, so I presume you travelled as far as Bentley with your father on his way to market and did you actually go from Bentley to Hadleigh alone? Tell me everything, there's a good pal.

Sleeping in my room are you? That's the proper place for you. Still, the room you had last time was better. I do hope you sleep well and enjoy yourself and, go home again looking better for the change.

So you've met Lyd and the babe (wonderful child). I expect Lyd gave you to understand there was no other child like it. How did you get on with Lyd? She was very masterful when we were kids and I used to fiercely resent her bossing me about.

Tell me the truth about the babe, you say it's a darling.

I do wish I could be with you, for there are some walks I want to take you for and goodness knows what. Tell me who you met and what you thought of them. I do hope you did not get badly bored. For there's no one to take you out. Do write and tell me what everything is like at the old place. You know I've not been there since a day or two after you were there with me.

All my love,
Tim

Tim's hut where he worked and slept. A food parcel from Pollie.

Sunday Afternoon
20.8.16

Dearest

    Time flies giving me few moments to write you. No. 48 dated 5th August was received a day or so ago. I do not quite make out where this Underwoods Hard is which offers you bathing facilities. Oh, I will teach you to swim. Sharks have been seen not far from here so I don't think I will risk such a dainty morsel as you in the water hereabouts, so we must wait until I come home again. I suppose Lucas got his camera back with the photo. How did it come out? Thanks for the organ recital programme. I should liked to have been there.

    Yes, Fred is very slang in his lang! I'm glad you liked him.

    I've got a hammock now in which I sleep and take my afternoon siesta. It's wondrous comfy and if you were here you would always be in it, for you are a real hammock girl. I slept in it for the first time last night and had quite a job to get up this morning. It was 5o/c before I could get on my feet. That will never do, it must be 4.30 at the latest or I must get down on the ground again.

    I am keeping well but have got a heap of washing and darning to do — I do detest such work.

    I hear from Jess and Len that Len has to go. It's unfortunate at such a time and Jess did so want him to be with her a bit longer. I hear that the young Oxborrow's birthday will probably be 20th August (today). I hope all will have many returns of the day.

    What a contrast — I picture in my mind B'sea today — so saintly and solemn, all in Sunday attire and Sunday airs and manners too. Then this place, the same yesterday, today and for ever and probably for thousands of years. We soldiers, British and Indians... Sikhs they are, a most picturesque race, bearing the name of "Paliala". Fine fellows they are, glistening white teeth, bronze skinned, black haired, they allow their hair to grow and tie it in a knot on the crown of their head. They are scrupulously clean and shampoo their hair almost every day down by the canal, then they let it dry in the blazing sun, most curious they look with their hair flowing down over their shoulders, they do not smoke, and they pray in public — a sentry coming off duty will kneel down for 15 mins prayer. They are much more devout than the English.

    I hope you are well and taking care of yourself and that Miss Lambert Jnr (what's her name) and PP (the boy) are well.

    All my Love
    Your Man

Troops from the Indian Brigade on landing at Kubri, with dows passing.

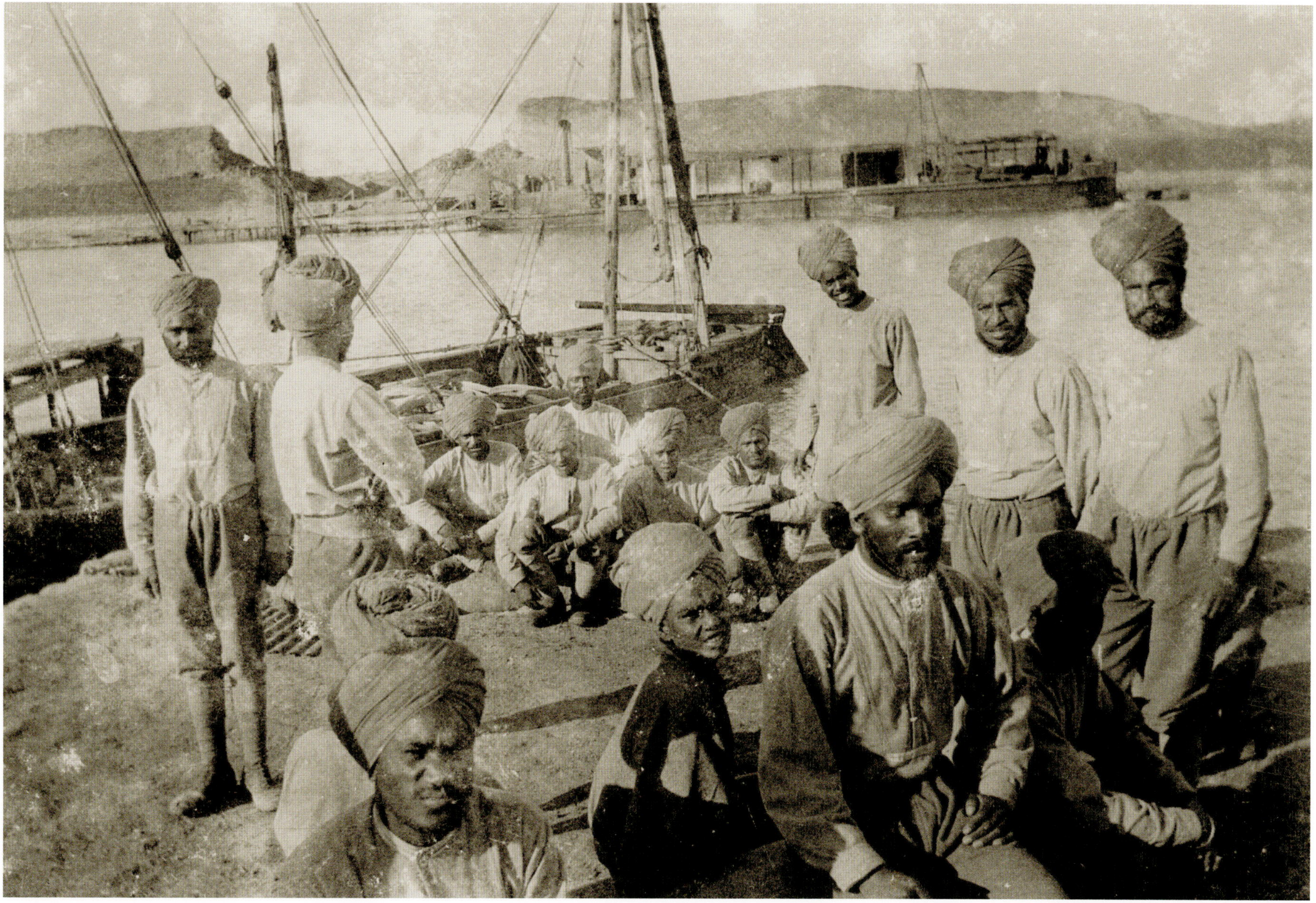

"An orderly of the Hyderabad Lancers. A Musselman by faith."

"A few of the men employed in work on the west side of Suez Canal."

"Platelayers. Worried my life out until I took their photos." Platelayers' trolly.

Friday 13.10.16

Dearest

    To be ceased in the middle of a very busy morning with a strong desire to write a love letter is rather annoying in one way, though rather pleasant in another. Fancy making a train wait while the station master writes a love letter! How ridiculous. This is such a morning as we have wandered off for a stroll, and I've been back to the station late, and now there is no one to say anything for me wandering your way again.

    The sun is brilliant and the air balmy. A soft gentle breeze drifts along sweet and fresh. It comes from the direction where lies England.

    Many thanks for the autumn tinted leaves and your letter dated 23rd September, and the parcel too. I was delighted with the contents - the sweets and the potted meat were a most welcome luxury and am truly thankful for them. The belt fits like a glove and will I hope contribute towards the goodness of my health. When I asked for it there was a bit of a cholera scare and we were all once again inoculated. But the scare has passed and by wearing the belt one guards against sickness. It's nice and warm and comfy, thanks.

    Your letter dear was very welcome. Indeed, you are getting on, cooking the joint, eh? And didn't you burn it?

    Quite exciting with the zepps - and quite near too. Am very glad you are still quite safe. Take care of yourself till I come back and then I'll take care of you (s'pose you'll allow me to?). Yes I like the curtains well, we shall have to have some of them, which room are they for? I really think they would be alright almost anywhere but still I must leave that to you, in fact the entire comfort of our home will depend upon you - just think what a lot you will have to do and all I've got to do is to drop down in a nice comfy chair and enjoy it all.

    Now my wife - how do you like being my wife? - I must conclude with apologies for a hastily scrawled letter and my fondest love.

    Your Man

Wednesday 18.10.16

Dearest

    Many thanks for No. 65 dated 1st October. The leaves you send are beautiful and much would I love to see the glorious colours of the trees in England now.

    This alteration of clocks must be a queer business. What with this and that and other things too we shall feel like fish out of water when we come home again or as if in a stranger land.

    How did you like the harvest thanksgiving service? I should certainly liked to have been with you. Well I suppose you've met the "young fellow" who's at the Lodge by now. What sort of a chap is he?

    A new navy suit eh? I should imagine it suits you. I might with an effort get down to the Hard with you on Sunday morning but really I would rather walk in less frequented ways, the very unconventional life here makes the Sunday morning hard harder, certainly less pleasant.

Please address me in future:
Light Railways,
"C" Subsection,
No. 1 Section
Canal Defences, EEF

    At present I'm reading "Candles in the Wind". I think you've read it. Did you like it? I heard form Blundell last week, he sent me his photo in Khaki and in civils too, also an etching — landscape view.

    Not many signs of autumn are to be found here, a few more clouds, a little colder at night time and evening but still warmer and brighter than midsummer in England during the day, and still no rain. We still go about in shirt sleeves and then find it hot, hotter than summer in England. No falling leaves here, no fog, no rain, few clouds, the sun risings are more beautiful than ever just now on account of the clouds.

    Did I tell you that I have the offer of a week's holiday in Alexandria? One wants plenty of "fuloos" (money) for that — there's plenty to see and it's beautiful by the sea in Camp. Really I should like to go and if a suitable opportunity occurs I shall do so. It was this time last year that I was in hospital there, perchance I might visit ward No. 15 and look up my old nurses and see who is the lucky occupant of my bed. Really it scarcely seems a year since I was there and sitting outside the hospital on the seafront writing to you and telling you how beautiful the scene was and how I was hoping to come to England.

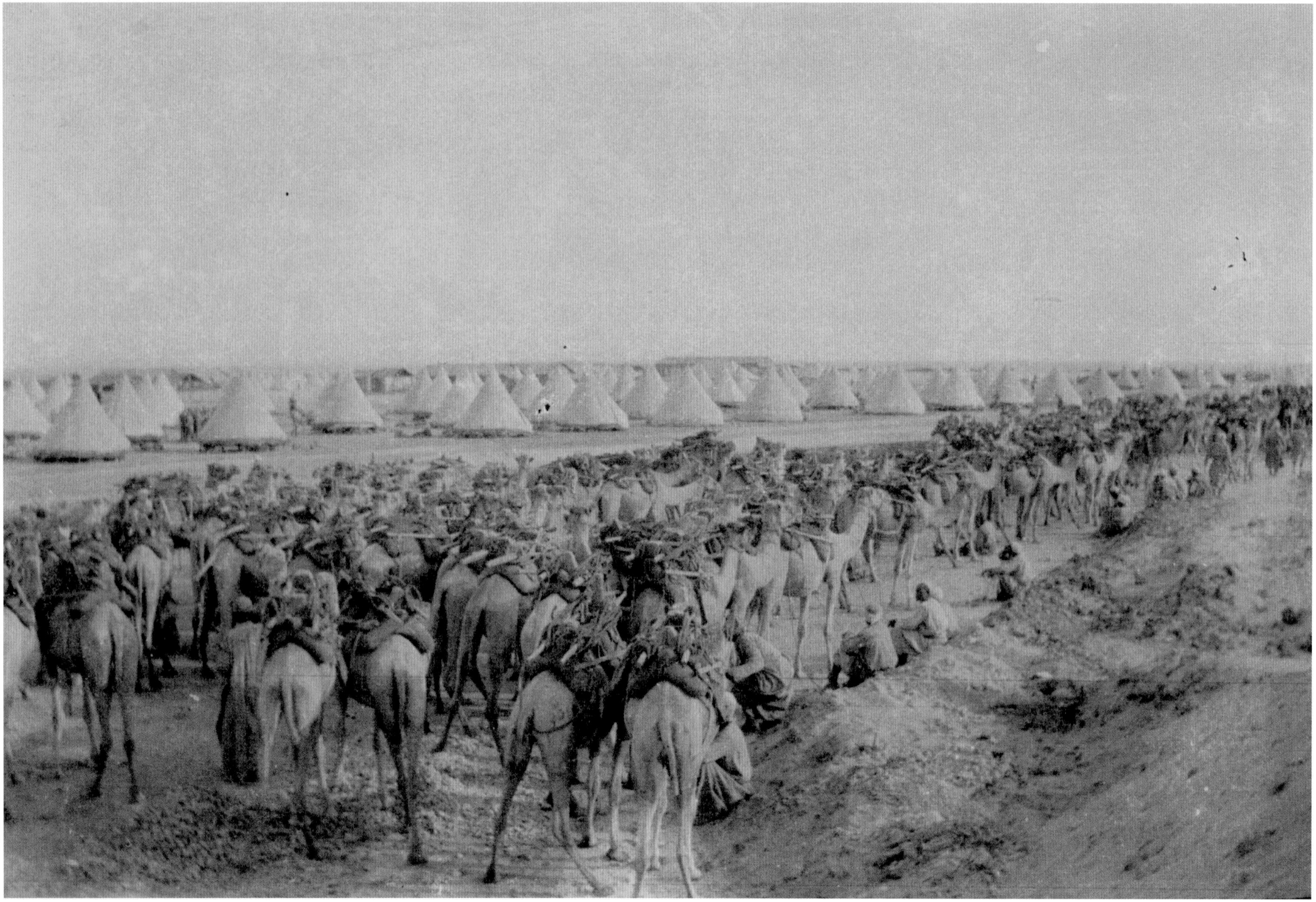

Left: "El Kubri East. Camel Corps waiting for orders."

Camels were the only practical means of transport over the soft sand of the Sinai Desert. 20,000 camels were employed by the EEF, rising to 35,000 by mid 1917. They were organised in companies of about 2,000, under Egyptian drivers.

"An unusual passenger just arrived here from the Desert – a lassoed camel, taken just outside my hut. By stretching his neck a trifle he could put his head through the window. The camel is lame."

Sitting on No. 2 bridge at El Kubri, catching a fish. Notice the ready-made sections of narrow gauge track stacked behind for extending the railway onto the bridge-head. Family tradition says he met the Prince of Wales (later Edward VIII) around this time.

Wednesday 20.12.1916

Darling

Your No. 80 arrived this evening in the parcel. The holly I was delighted to get, the sweets too and your letter is most cheering and welcome this evening — the end of a very hot and vexatious day. News came to us of another mail having been sunk. Oh, it is most annoying! Everything seems to be torpedoed. It was rumoured that our Xmas mail home was sunk but I never saw or heard confirmation of it, and today we learn that letters posted between 1 and 4 Dec have gone down. I believe my letters have been correctly numbered lately — are there any missing?

Dearest, you presented me with so many volumes of RLS [Robert Louis Stephenson] on my last birthday I really cannot remember which of his books I have, yet I know there are books of his letters I have not read. One — "Vailima Letters" I have. How I wish his skill and gift of writing were mine — what a letter I could write you! They would really be worth reading many times and would give you a wealth of happiness and contentment. RLS was the favourite author of Lt Hopkins who used to be in charge here. Often we would discuss his books. It occurs to me you are desirous of sending me a book upon the occasion of my birthday. I'm sorry Poll dear, I can think of no book in particular, neither can I think of anything else. Truly I would give you a hint if I could but can think of nothing. The greatest gift of all I have — your love. It is ever a source of comfort, sympathy and happiness to me. Often I turn to it, when troubles and limitations beset me, when all seems black and my heart is heavy, I turn to your love and find it the best friend and treasure I have, and go on my way again strengthened and cheered. Often in fancy I come to you thus, as in fact I shall come to you in the sweet by and by when the day is done. I'm glad we understand each other so well for it enables us to discuss and talk things over with a greater freedom, and we shall have much to talk over when we meet again.

I believe I told you and often bewailed to you my scanty pay when I was in the Suffolks. You will be pleased to hear that my transfer to the Engineers more than doubled that miserable 1/3 per day. Today the rate on my pay book is 2/10 per day as a sapper. Shortly I hope to get the additional -/8 per day to which I am entitled as a 2nd corporal making a total of 3/6. My officer is trying to get further promotion for me but it is difficult — my transfer from the Infantry is so recent and my rank in the Infantry so low so perhaps I shall have to wait and content myself. How much do you calculate I can save per month? I fancy I've nearly got enough for 1 leaf of that gateleg.

Tim in his hammock in his shed.

This date last year I was released from gaol at Birmingham. This day last year you flew into my arms for the first time since we parted in July — engaged. Today I'm where I was in April last and it's as hot today as it was in July or August, scorching, and has been so for some days, quite a heat wave. Here we are within 5 days of Christmas and it's 80° in the shade.

I hope you are happy and well. Thanks very much for the parcel of sweets and holly, I was very glad to receive it.

My love as always
Your Man

Note requesting permission to stay up late. Permission granted.

"Breakfast Christmas morning 1916. Note the vacant place bottom left hand corner."

Tim seems to have been in charge of Xmas day celebrations. Here is his itinerary, including the names of the people and what songs they are going to sing. He even ticked them off as they sung them. On the reverse side of the paper are more acts, but it becomes illegible as the night wears on…

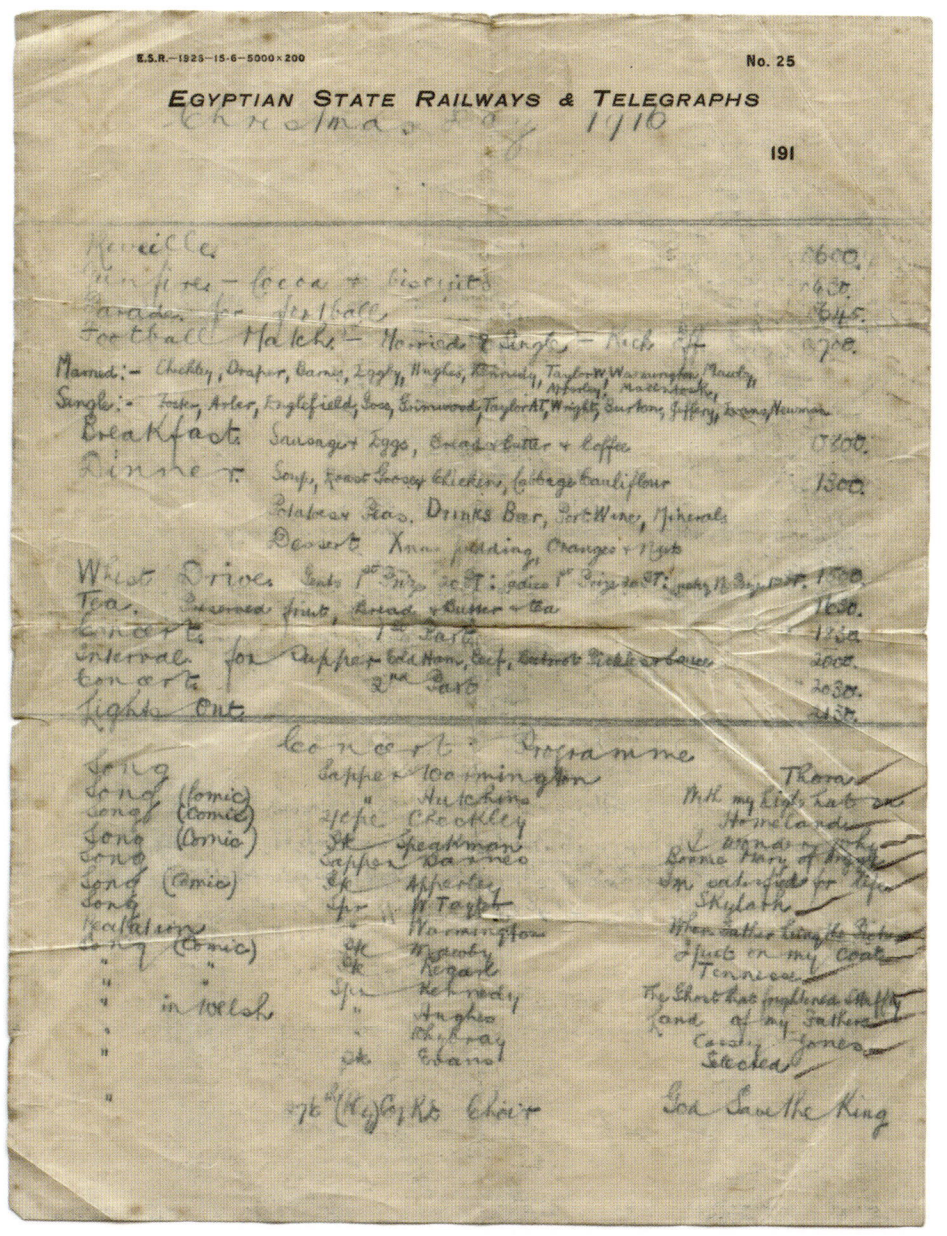

Christmas Day 1916

Reveille	6.00
Gunfire - Cocoa & biscuit	6.30
Parade for football	6.45
Football match. Married v Single. Kick off	7.00

Married:	Checkley, Draper, Barnes, Eggby, Hughes, Kennedy, Taylor W, Warmington, Mawby, Apperley, Macintosh.
Single:	Foster, Arber, Englefield, Goss, Grimwood, Taylor AT, Wright, Burton, Jeffery, Evans, Newman.

Breakfast:	Sausage & Eggs, Bread & butter, Coffee.	08.00
Dinner:	Soup, Roast Goose, Chicken, Cabbage, Cauliflower, Potatoes & Peas. Drinks: Beer, Port Wine, Minerals	13.00
Desert:	Xmas pudding, Oranges & Nuts	

Whist Drive.	15.00
Tea. Preserved Fruit, Bread & Butter & Tea	16.30
Concert 1st Part	18.30
Interval for Supper: Cold Ham, Beef, Beetroot Pickles & Sauce	20.00
Concert 2nd Part	20.30
Lights out	21.30

Concert Programme

Song	Sapper Warmington	"Thora"
Song (Comic)	Sapper Hutchins	"With my high hat on"
Song (Comic)	2/Cpl Checkley	"Homelands"
Song (Comic)	Pte Speakman	"I wonder why"
Song	Sapper Barnes	"Bonnie Mary of Argyl"
Song (Comic)	Pte Apperley	"I'm satisfied for Life"
Song	Sapper W Taylor	"Skylath"
Recitation	Sapper Warmington	"When Father hung the Picture"
Song (Comic)	Pte Mawby	"I Put on my Coat"
Song (Comic)	Pte Regail	'Tennessee"
Song (Comic)	Spr Kennedy	"The Ghost that Frightened Shaffly
Song in Welsh	Spr Hughes	""Land of my Fathers"
Song	Spr Whybray	"Selected"
Song	Pte Evans	"Casey Jones"
Song	276th (Rly) Coy Rt Choir	"God Save the King"

Trams in Alexandria, photographic adverts in the *Palestine News*.

Suez station and goods yard. Cinema flyer and train tickets.

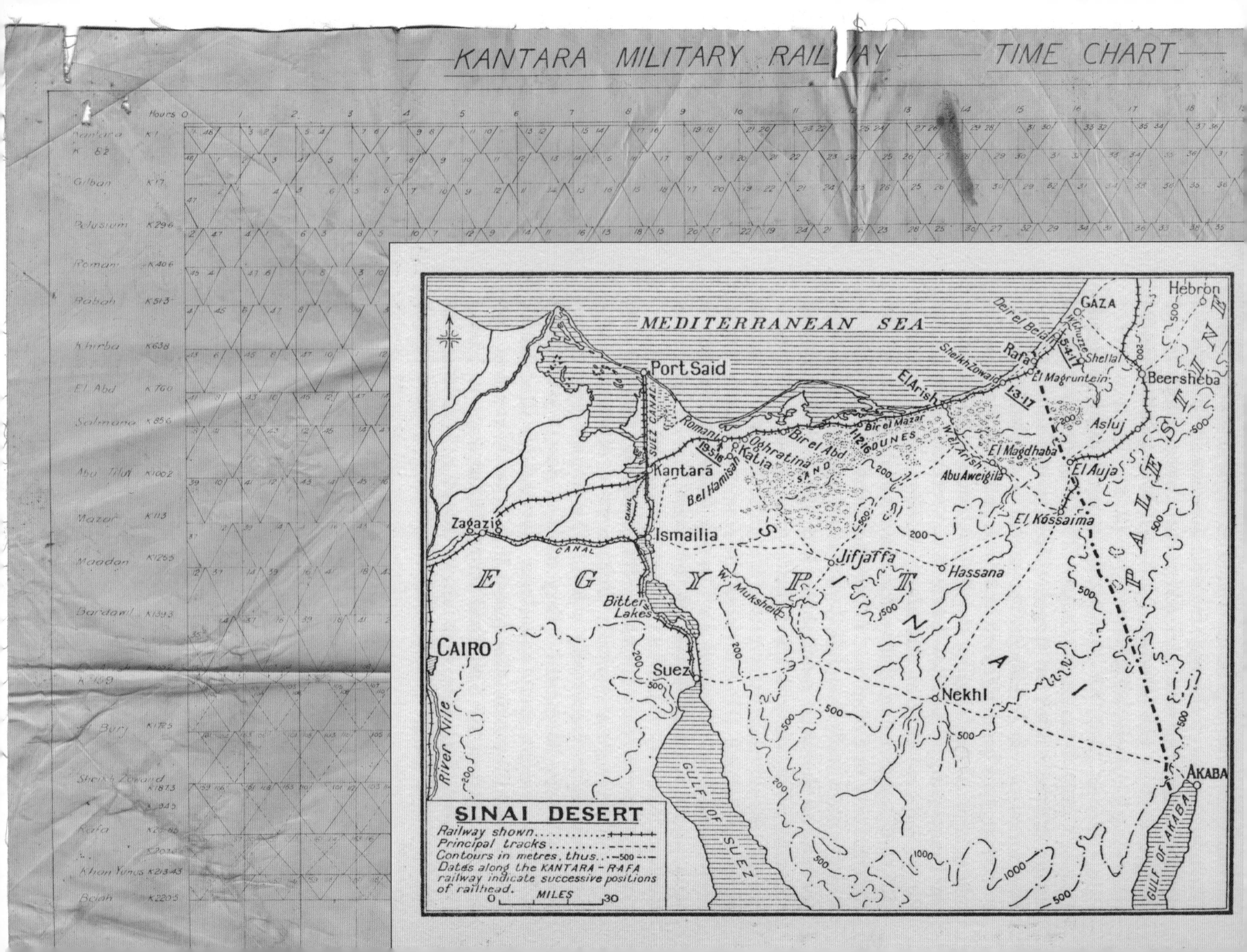

KANTARA MILITARY RAILWAY — TIME CHART

SINAI DESERT

Railway shown ....
Principal tracks ....
Contours in metres, thus.. —500—
Dates along the KANTARA–RAFA
railway indicate successive positions
of railhead.

MILES 30

1917: THE SINAI MILITARY RAILWAY

In February 1916 the EEF began building the standard gauge (4ft 8½ inches) Sinai Military Railway from Kantara on the Suez Canal, across Sinai towards Romani and up the coast. Alongside was a water pipeline to supply the steam engines and provide drinking water. They were laid and operated by the Royal Engineers, and in April 1916 the 276th Railway Company was formed to operate the railway, to which Tim transferred at the end of 1916 (with a pay rise). The railway advanced as the EEF took more land to the north from the Turks: Romani May 1916, El Arish January 1917 and Rafa March 1917. In March 1917 the 53rd Railway Troop Company and the 274th and 276th Railway Companies were combined to form the Railway Operating Division (ROD). A branch line was built from Rafa to Beersheba which was captured in October 1917. Gaza was won in November 1917, shortly followed by Jaffa further up the coast. The British entered Jerusalem on December 11, 1917.

Tim left Kubri at some time in 1917 and followed the line as it progressed along the coast across the desert. He didn't put his location on these letters, but there are photos of Bardawil (March and April), El Arish and El Imara. There are no letters from him between May and October, after which we find him convalescing in Alexandria.

Stations on the Kantara line to Gaza (147 miles)

Kantara East	Romani	Ab Tilul	El Burg
Gilban	Rabah	Mazar	Shein Londid
Pelusium	Khirba	Mardan	**Rafa** (March '17)
Romani (May '16)	El Abd	Bardawil	Khan Yunus
Mahamdiya (June)	Salmana	**El Arish** (Jan '17)	Benisela
			Gaza (Nov '17)

Left: Kantara Military Railway Time Chart.
Inset: Map from *The Palestine Campaigns* by Lieut-General Sir Archibald P Wavell CMG MC. London Constable and Co Ltd. 3rd ed. 1941.

"The Railway depot at Kantara East."

Wednesday 3.1.1917

Dearest

    Whenever I set me down to write to anyone, no matter to whom it may be, I almost invariably start off by addressing myself to you. Often have I sat down with the intention of writing to Jess or Mother or Blundell, but have written you instead. I have several letters from various correspondents to answer, I sat down just now to answer these letters and thus, as usual, I am writing to you. Not that I have much to write excepting to sing the same song to you — I fear that song is often flat and very much out of tune, nevertheless it pleases me to sing it and so long as you can endure it I am content.

    Well sweetheart, a funnelless steamer passed here this morning — the first I've seen for many a week without a funnel — but it was sailing northward and was flying the Danish flag and on it were the words SIAM of KOBENHAVN and no Jack Ainger was on board. I fear I have missed him.

    Ever since Xmas the weather has been much colder — cloudy and rainy too. How is the weather at B'sea? Cold I suppose.

"Sick Camel Traffic. Bardawil."

Saturday Evening 6.1.1917

My Wife

    It's a most glorious moonlit night, snowy white clouds bespatter a beautiful starry sky and the air is sweet and fresh. I've been very happy all day long for I received your No. 86 of 11th December this morning. The mails seem to be very irregular and late just now. There was a mail due in on 5th January but it has not come yet. You seem to be happy dear — I'm glad. Oh, how delighted I was to get your letter!

    You tell me that you live in a world of your own — you dream and are happy when dreaming of our home. It pleases me well that you should find happiness in dreaming of our home. I'm the same Poll dear, and it's there I find my happiness too. Often I take your hand and wander in the land of dreams. We go for some most delightful trips, how dear and near you are then. So close I hold you, and you nestle close to me too, your hair falling over my shoulder and your lips seeking mine. Sometimes you take me from room to room and show me our home, at other times we sit by the firelight and talk — you on the hearth rug, myself in the armchair. Sometimes you talk to me about Richard [Tim and Poll's imagined children: Richard and Mollie]. Always we are very very happy.

    I often thank heaven for the truth and constancy of your love.

    I'm glad you keep busy — nothing like plenty of work for contentment and peace. In addition to passing the time away quicker it will help to make you a good strong and capable little wife, which qualities are a great asset to happiness in a little home like ours.

    Now little wife it's late and I've had a busy day and have the prospect of another busy one tomorrow and shall be glad to lie down and sleep.

God bless and take care of you my wife — my love,
Ever your Man

I think I have told you to address me to:
276th (Rly) Coy, Royal Engineers etc.

Standard gauge Manning Wardle 0-4-0 petrol loco on armoured train.

Thursday 1.2.1917

Little Wife

Whenever did I write you last? The mail seems few and far between.

Well, I've been somewhat occupied the last few days chiefly on account of 48 hours leave from which I have just returned. It was 1o/c last night when I dragged two weary feet over the sand from the station at Kubri West after a long journey and a very welcome change. I went on a visit to my brother Fred whose battalion is on the move today — he was about 2½ hours train journey from me and is going much further away. So I got leave to give him a look. Returning yesterday afternoon I got into a wrong train at 2o/c. At 3.30 I was stranded at Tel-el-Kebir until 7pm. Tel-el-Kebir is a most desolate place — a native hovel or two and a cemetery. The cemetery is a small enclosure shaded with palms and other trees and marks the resting place of those who fell in the battle of Tel-el-Kebir and Kasasnas in August 1882. Rude crosses tell that "One Unknown" or "Two Unknown" lie beneath, and here and there a stone with a name and date and a word or two to tell that he was killed in the midnight charge. Then a few who died in the present war are resting there too — all soldiers. And from 1882 to 1915 none were buried. It was very interesting and I took one or two photos which if successful I will send you. At 7o/c I took train again and it was 1am before I reached my bed — 6 long weary hours and nothing but desert.

Standing in a shop at Ismailia I met Clarence Kirby who is in Noble Eagle's crush. I had not seen him since I met him on Gallipoli with Noble Eagle. He looked very well.

I went in a silk shop at Ismailia — the fellow wanted to sell me a suit of ladies pyjamas — silk ones. Very flimsy they seemed, still very pretty and nice, but I didn't think them exactly the colour or size and "mafeesh fuloos" (my money was low) so I did not do a deal. They've got heaps of pretty things at these shops but I'm very ignorant as to their quality or value.

All my Love,
Your Sweetheart

Foster brothers Fred and Tim.
Fred in his dugout.

Midday Friday 9.3.17

Dearest

I think the flies are just about in the height of their glory. I've got that mosquito net you fitted and helped to make. I am now wearing it, and mighty thankful I am for it for this morning is scorching hot and the flies are biting and worrying in swarms. They actually settle on the lather when one shaves just for a drop of moisture, and almost snatch the food from one's hand before one's mouth can close upon it. Jam — why, one has to lift the cover of the tin, snatch a thimbleful out, daub it on the bread and snap it up all in an instant, and then perhaps lose half of it and swallow a host of flies with the other half.

Now dear, if you want me to survive this war — if you truly love me and do not want the flies to completely consume me, and if you would like to get your hand in, just send me a piece more muslin or netting — it need not be as fine mesh as that which Mother sent me last year, for the flies which pester us here are just ordinary house flies — gone mad! And so the ordinary netting would do, and would in fact do better, for the very fine stuff that Mother used does not admit the air as freely as the larger mesh stuff would. I do not want it made up in the fashion Mother made it last summer — that style is alright for walking about or sitting down, what I want is something to keep the flies off me when lying down. Often my hours of rest and sleep fall during the day, then it is that I want a plain length of netting to stretch over me. If it's not too expensive I should be glad of a yard or two sufficiently wide not only to cover me but to leave a good margin say 1½ to 2 yards wide and 3 yards long. Then it would be very convenient if tape were threaded along each side so that it could be drawn fast and tied, thus I may sleep secure from all harm and the netting around and about me having been blessed and sent by your own hands will symbolise to me your presence as a guardian angel watching over me.

I have no idea as to the price of this material, neither have I any idea as to the state of your exchequer, but if the stuff is dear and you are stoney then "malish" — (never mind).

Now I have a great desire to comfort and cheer you. Still I was jolly glad to get your letter last night and very sorry that you were so sad. I too have sometimes been miserable by having no letter from you for so long a time and can sympathise with you when you tell me no letter has reached you for nearly 2 weeks. I sincerely hope you received letters from me shortly after and that long since your heart was light and your face bright again.

Sunday Morning
29.4.17

Dear

What a little and common word with which to greet one's best beloved. How often used as a mere matter of form and courtesy yet when I greet you with that word how different its meaning. Picture in your fancy a little home, the day's work done, the fire is burning bright and cheerful in the cosiest of rooms, tea is already laid out upon the table — for two. A footstep outside, a door opens and is hardly closed before the husband has his wife in his arms and into her ear he murmurs that word "dear". Often have I endeavoured to give full expression to all that I feel and have in my heart for you but never once have I succeeded. The deepest, the truest, the strongest and the best cannot be written, so it is in that which is not written lies the best and the fullness of my love for you. The love that lies on the surface does not lie in the depths. Words are superficial, true love itself is profound — deep — and I hope that beneath my words you are able to perceive the profound depths of my love for you. I hope so because I feel sure that such perception will ever be a source of happiness for you.

My poor little pal — my wife — I am very sorry indeed yet I am glad you should write and tell me how sad you are. No matter how you feel dear you know you always have my sympathy. Please — my darling — do not ever think I shall be vexed because you write me short and sad — it hurts me that you should turn to me for sympathy and doubt whether or not I shall give it to you. Do not doubt me Poll. When I read your letter I was sad because you were, but I was happy because I know it was your love for me that made it possible for you to write and tell me of your sadness. Poll, little wife, we share joys and sorrows too — halves in everything — don't be greedy and keep all your sorrows to yourself, there's nothing like sorrow for cement.

Now Mrs Tim Foster, next time you write short and sorry do not ask me to forgive you! Our love goes deeper than mere praise or blame, punishment or forgiveness. It's a wonderful, immortal mystery in which our two lives are merged into one. We stand or fall together, your righteousness as well as your unrighteousness is mine and mine yours. Do you think our united strength can bear the unrighteousness of us both? This conception of love guides my footsteps and you — my lover — help, guide and strengthen me and so your love makes me a better man than I should otherwise be.

I hear little pal that this may be very boresome, it's a weakness of mine to reason and ramble, but you do not mind, I've got lots of other things to do and letters to write, but I like writing you best. Why? Because I can ramble and write as I please and know I shall be understood. Even if my reasoning and writing is tangled and not understandable, and even boresome, I shall be understood and find that which I want — sympathy. Of all things I prefer to write to you.

I hope long before you get this letter your downheartedness will have passed. Perchance some letter of mine reached you about that time and cheered you. You wish it were possible to come to you dear — do you know what I should do — or want to do — upon meeting you — I should want to give you a hearty handshake, such a handshake as I should give to my dearest pal and say "good girl" then almost before this was done up would jump my love and my arms would be round you and my kisses upon your lips.

Write and tell me everything and all the news.
All my Love
And Always,
Your Man

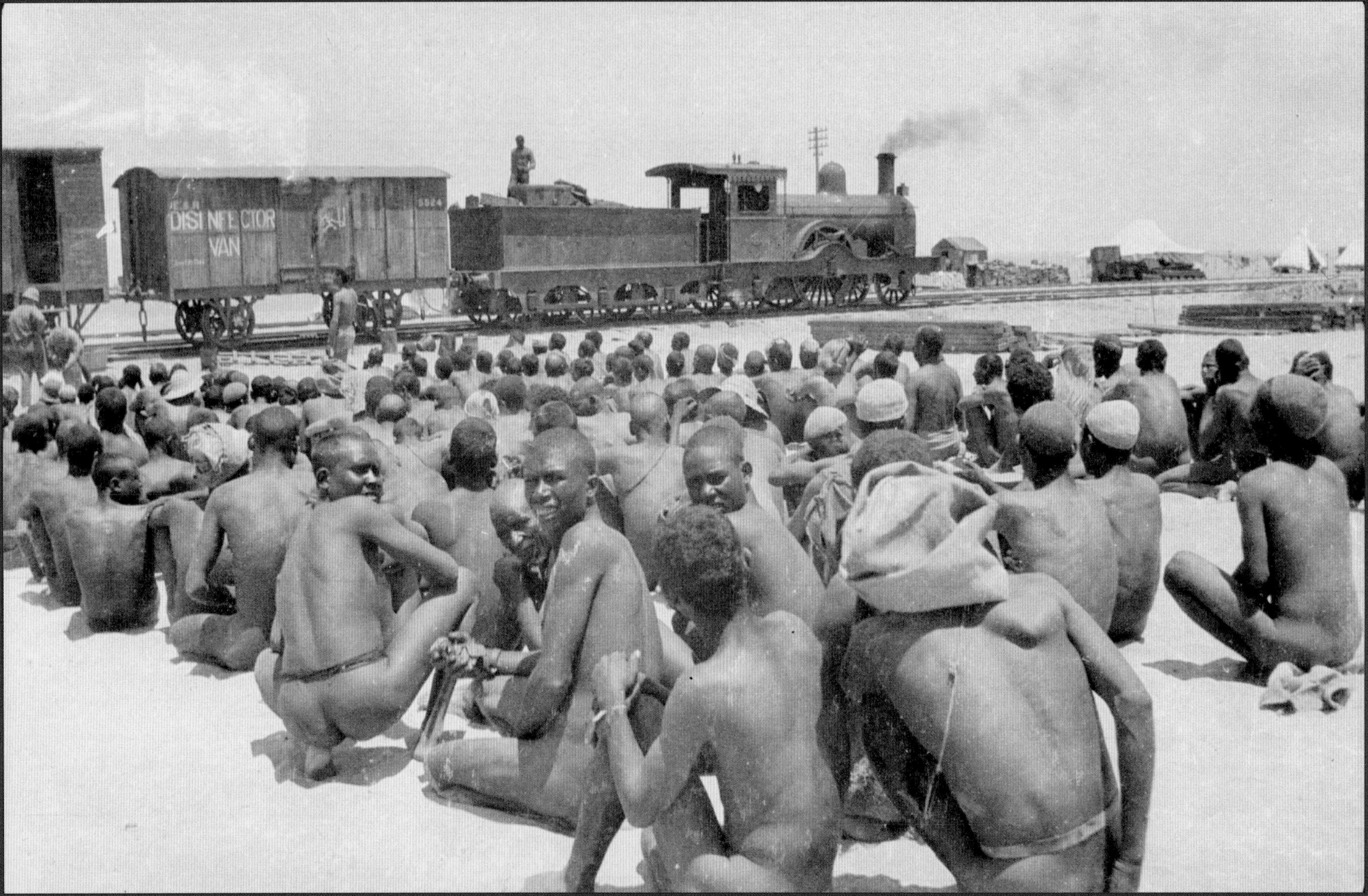
DISINFECTOR
VAN

Recruits for the Egyptian Labour Corps (ELC).

Egyptian labour was invaluable in the desert. These men usually came from poor Egyptian and Sudanese villages and the daily rate and rations were very attractive. The recruit was issued with blankets and an overcoat and at the service depot he was disinfected, given clothes and equipment. Eventually there were about 100,000 in the ELC. After a while the British began to turn a blind eye to the use of press gangs, and unrest and riots followed in 1919.

Thresh high-pressure steam disinfecting machines were transported out to the desert to sterilise uniforms. They were worked by steam from the engine, and consisted of two vans capable of cleaning up to 300 kits. Also required were antiseptic baths for the recruits due to body parasites and several cholera and scabies outbreaks.

"General view of Bardawil from top of sand hill overlooking the sea. In distance train is seen standing on station."

"Foreman of our native platelayers, commonly known as Reis (Rice)."

Moving a platelayers' shed.

"Platelayers outside their shack near the 4 Kilo post on the Kubri line."

"Egyptian Army recruits. Platelayers on Palestine Military Railway."

"Bardawil Station. Egypt-Palestine Military Railway"

Red Cross train.

Two Baldwin locos from the Egyptian State Railway pulling an ambulance train

Red Cross train.

Collision at Bardawil 10.4.1917
See next page for documentation: "Ambulance 32 collided at west leg Points with Ration 7 at 0946"

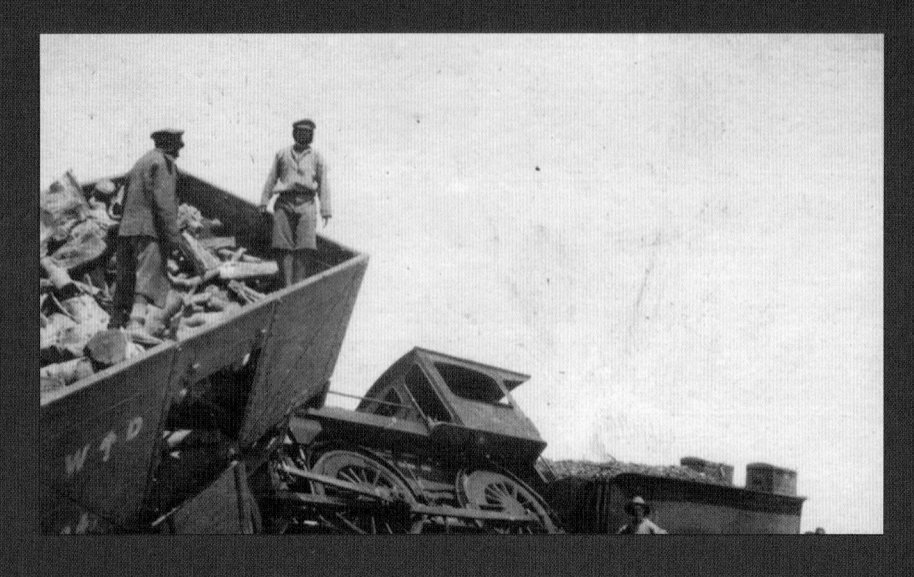

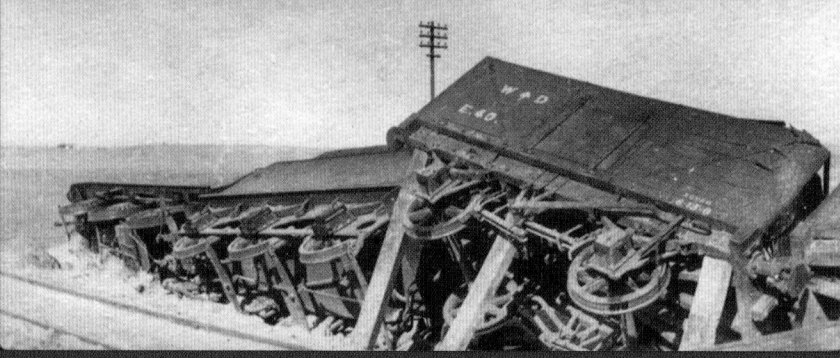

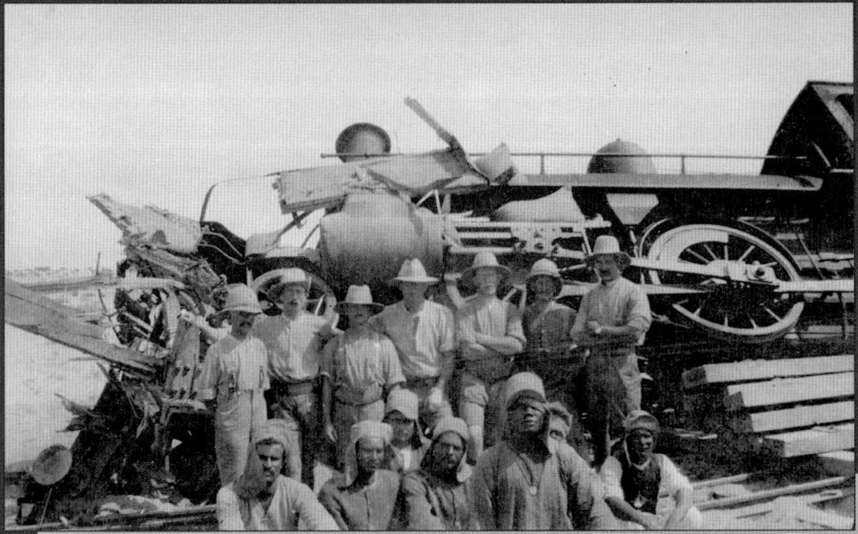

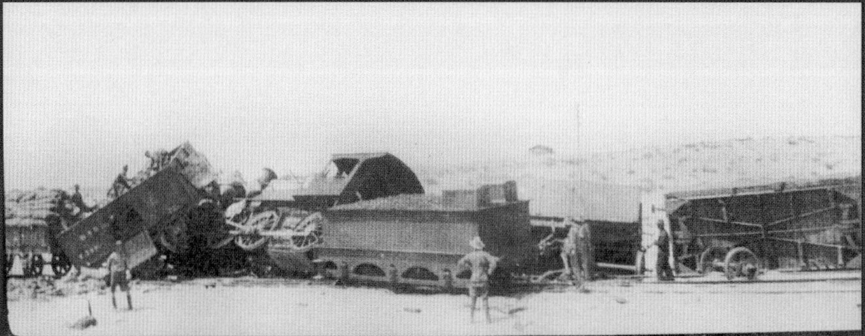

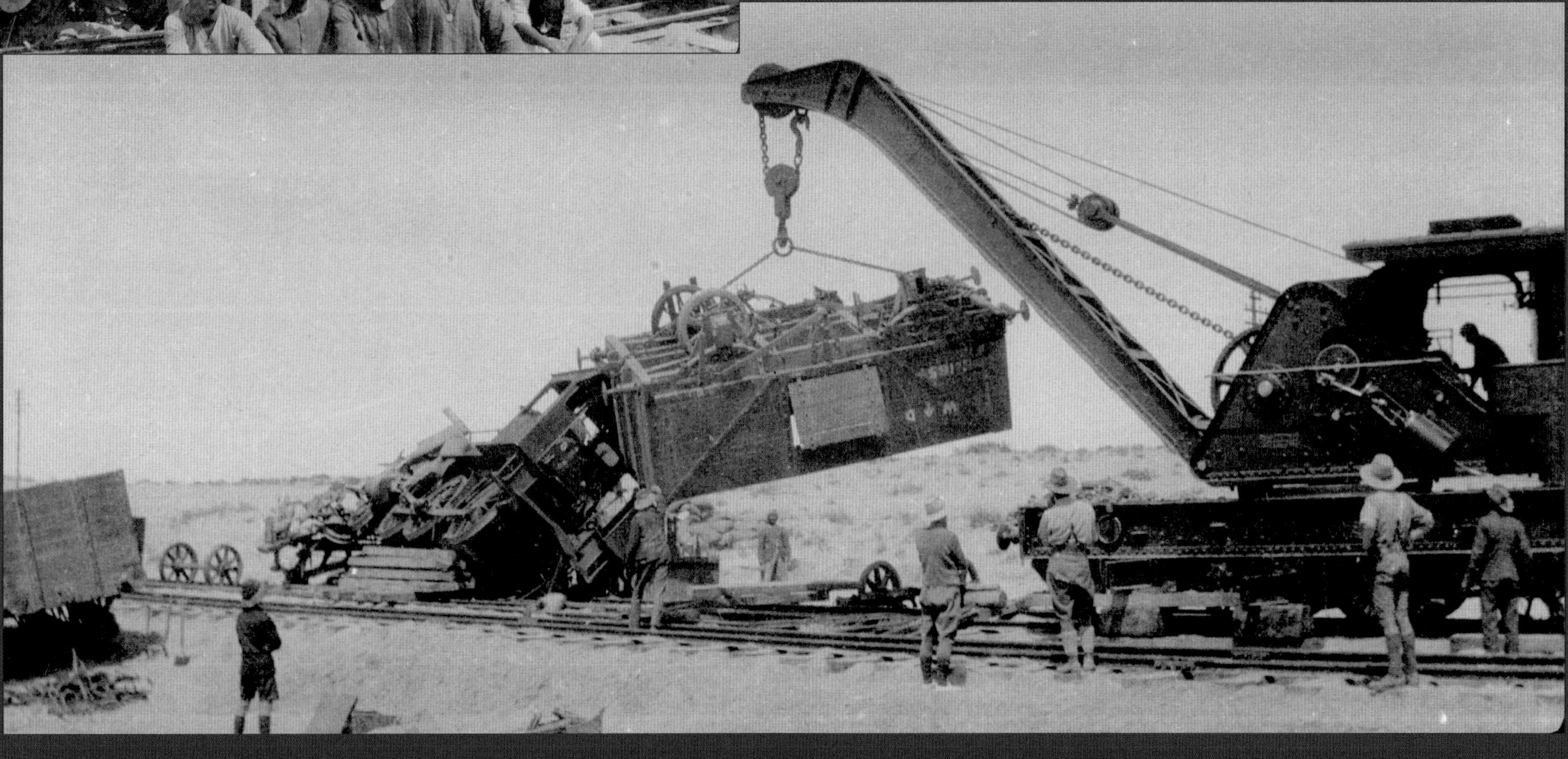

EGYPTIAN STATE RAILWAYS — TRAFFIC DEPARTMENT

TRAINS REGISTER — Station: Bardawil Date: 10th April 1917 Tuesday

UP TRAINS — القطارات الطالعه

(From 12.1 a.m. to 11.59 p.m.)

Description	No. of engine	Asked for by station in rear	Given to station in rear (Uncond / Cond)	Train entering Section Code recd from rear	Train out of Section Code sent to rear	Asked for from station in advance	Received from station in advance (Uncond / Cond)	Train entering Section Code sent to advance	Train out of Section recd from advance	Time given/recd Blocking Back	Obstruction removed	Booked Arrival	Booked Departure	Actual Arrival	Actual Departure	Remarks
Lel gos	14	0005	0005	0015	0055	0719	0719	0725	0758							
			Working No 37 Drawing No 37 over													
Spl Etos	16	0120	0120	0128	0159	0231	0231	0303	0310							Wtg arrl No 39
Do	12	0242	0242	0252	0322	0407	0407	0437								Wtg arrl No 41
Spl Stos	26	0555	0555	0606	0638	0622	0624	0641								
Do	28	0716	0716	0725	0758	0743	0743	0802	0834							
Amb	30	0840	0840	Cancelled at 0900												
Amb	32	0915	0915	0923	0945											
		Collided at west-end pts with Return No 7 at 0946														
Rick down		1034	1034	1040	1111											
Do		1607	1607	1610	1643											
Spl Etos	6	2055	2055	2100	2130	2120	2120	2138	2208							
Do	12	2323	2323	2340	0013	2355	2355	0016	0043							

DOWN TRAINS No. 6

سكة حديد الحكومة المصرية — نمرة ٧٠١٥ا

كشف قيد القطارات بمحطة _____ بتاريخ _____

القطارات النازلة (من الساعة ١٢ دقيقة ١ صباحاً الى الساعة ١١ دقيقة ٥٩ مساء)

No. T.L. 77

Description	No.	No. of engine	"IS LINE CLEAR" Given to station in rear		"Train entering Section" Code received from station in rear	"Train out of Section" Code sent to station in rear	"IS LINE CLEAR" Received from station in advance		"Train entering Section" Code received from station in advance	"Train out of Section" Code sent to station in advance	"BLOCKING BACK" CODE Time given or received	Obstruction removed	BOOKED TIME OF ARRIVAL	BOOKED TIME OF DEPARTURE	ACTUAL TIME OF ARRIVAL	ACTUAL TIME OF DEPARTURE	Minutes late arrival	Minutes late departure	LEVEL CROSSING GATES Time Opened	Time Closed
			Asked for by station in rear	Uncond. / Cond.			Asked for from station in advance	Uncond. / Cond.					H. M.	H. M.	H. M.	H. M.				
arr	31		0045	0045 0052	0110		0110			cancelled at 0120									
						0159	0159												
ation	39		0158 0158	0200	0231 0322		0322 0324 0402												
			10tg arrl No. 72																
	41		0310	0310 0330	0402 0402		0402 0404 0442												
					0503 0450		0450 0507 0558												
	1		0347 0347	0532	0644 0758		0758 0800 1832												
			waiting arrival of no 26 + 28 tbe Etro																
ation	4		0905 0915	0922			crossing west pt at 0946												
			collided at meeting point with mll no 32 at 0946																
eng			1155	1155 1206	arrived at 1230														
tracks					1318 1318		1330 1330 1405												
					1259		1259 1304 1345												
eng			1409	1409 1410	arrived at 1440 at obstruction														
			Returning		1514 1514		1519 1544												

Electric Staff instrument made by Webb & Thompson for signalling and controlling the movement of trains.

Tim's rule book.

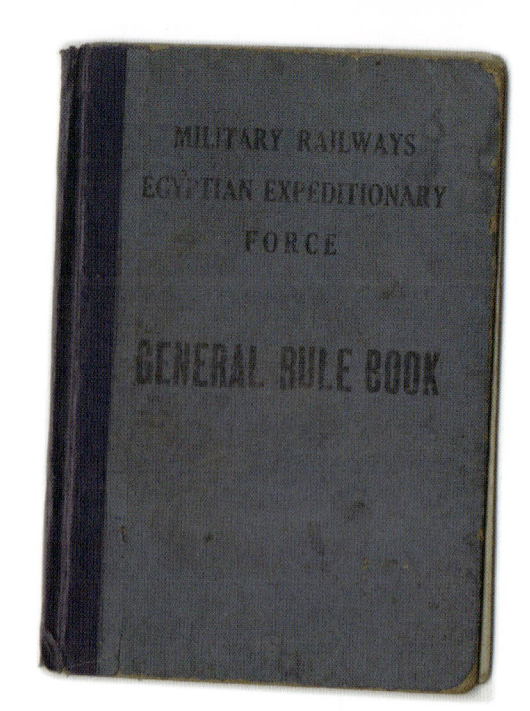

— 36 —

METHOD OF WORKING THE STAFF APPARATUS.

1. No Driver may enter any section between two Staff Stations without the Staff for the section being in his possession. When the line is clear between any two Staff Stations, a Staff can be obtained from the instrument at **either** end, but as soon as a Staff has been taken from either of the pair of instruments controlling the section, no other Staff can be obtained from either of them until the first Staff has been replaced in either of the instruments controlling the section.

2. This is effected by Electrical interlocking between the instruments, and as a result it is impossible for two trains to be in any one section at the same time, **provided the driver does not leave without the Staff for the section.**

3. The method of working is as follows :—
Suppose "A" and "B" represent two consecutive Staff Stations. When a train is at station "A" and requires to go to station "B" or when "Train entering station" is coded from the Staff Station in the rear for a train which requires to go to "B", the signalman at "A", provided he has received

— 37 —

the "Train out of section" Code for the previous train, and permission has not been given for a train to approach in the opposite direction, must give one ring to call the attention of "B". The signalman at "B" will reply by one ring. The signalman at "A" must then give the proper "Is Line Clear" Code (according to the description of train), and if the line at "B", upon which the approaching train is to run, is clear and free of all obstruction, the signalman must acknowledge the Code by repetition, thus giving the necessary permission for the train to leave "A".

When pressing down the key for the last time, in acknowledging the "Is Line Clear" Code, the signalman at "B" must hold it down (which will cause the Electric Needle at both "A" and "B" to fall from its upright to a slanting position) and the key must be held down until the Electric Needle returns to its upright position. The movement of the Electric Needle to a slanting position will tell the signalman at "A" that he can take a staff out, which he will proceed to do, after turning his Right-hand indicator to "For Staff", and the actual withdrawal of the Staff will mechanically replace his Right-hand indicator to "For Bell". (N.B. Certain types are not fitted

Sunday Night 20.5.17

My Darling

For 24 hours your two letters which reached me this afternoon were held up a few miles away by a sand storm. Never may I experience another such storm, it was a south wind which came from over the Arabian desert like hot air belching out of a blast furnace, and the heat and sand was cruel. But is passed now, your letters have come and I'm a happy man, dear old Poll. I feel as though I want to hug you ever so much, to kiss you would be the sweetest bliss imaginable.

Now my little wife — I love to call you that — mine — you are mine. I have two letters to answer, one posted on 30th April and one on 2nd May so if you will come with me I will take you to a secluded spot where the sand is soft, warm and silvery within sound of the rolling sea, and there we will talk about that wonderful scheme of yours of which you say "you will think my awfully mad, but I've thought so much about it I simply had to pass it on to you". Oh, how I want to hold you in my arms and kiss you! You are a dear and not a bit mad — the idea I mean.

No I do not think it will ever be used as a ref room again [the railway refreshment room, Brightlingsea station], in fact in the early months of the war Mr Ruffell told me that they were talking about making it into a residence for him, but it never came off, and I'm not sure that he was keen on it. Your idea would not be so impossible as you may think and I can tell you your brainy idea has fairly caught on with me. I believe I'm going to write a terrible long letter — for I'm going right through the house with you dear. I agree with you the bathroom is jolly handy and would do away with the necessity for wash stands in the bedrooms. I agree the kitchen is a place of great convenience.

Now dear just follow me. There's the kitchen, the small room and the larger room downstairs. How would it be to have breakfast, dinner and tea in the kitchen on all ordinary summer's weekdays. Suppose the table is beneath the window, we could have the window slightly open letting in a gleam of sunshine and a most pleasant breath of fresh air. In this way you would have everything handy, you could boil the eggs, make the tea, and have everything handy and not have to keep running into other rooms. Everything for cooking meals and clearing up right on the spot, and if I remember rightly it's very pleasant in the winter. I would suggest we spend the greater part of our time in the small room — you could cook in the kitchen and then we could have a fire in the small room and be nice and warm in the bleakest of weather, and no matter how dreary it might be outside we should be wonderfully snug. We could have one easy chair and one small wicker, a thick rug, a small table, just big enough for "we two" for meals. Then in summer or winter when we had company we would use the big room which could be furnished as you suggest with the best of stuff and suitable for a dining and sitting room. Here we could have a settee with two movable ends, broad soft and deep, a great big easy chair, bookcase, and if we were lucky a piano, a good table and lots of other things.

Now how does that suit, wife mine? Now come upstairs with me and see what we can do. If I remember rightly there are two small bedrooms and one slightly different with two or three windows. Well how would it be if we fixed up the one facing the Lower Park Road for ourselves, could you get your dressing table in it? And a wardrobe too? Really it's not a bad view from the window and by having the windows open we could breathe the sweetness of the air, hear the birds sing and whistle early in the spring mornings, and should get a bit of the sunrise. Then the next small room could be fitted up for chance visitors, leaving the other room which faces the river. Now how would it be if we were to fix that room up for summer afternoons and evenings. We could have the windows wide open and the sun would shine in through them and the sweet scented breeze too would drift in. We could have tea there, and there we could sit on summer evenings and talk, read and write and together watch the sun set as we used to up the line. I want to pack up and come at once and have all these things come to pass. Oh, what a treat to have a good bath in a real bath with plenty of water to rub and dry oneself on a glorious clean towel, to wear soft and clean shirts, to sleep in a real bed with real clean sheets, to sit and eat decent food at a table without fear of its collapsing any moment. Oh, what a heaven! And you dear — to have you all for my own to love and cherish, my wife and my pal.

Your curtain arrangement for the large room will do A1.

That cup of something hot at 11am on a cold day will be very nice, but that tiny kiss you promise will be heavenly. Poll dear, what would I not give to have your kisses now, never have I longed so much, never have you been dearer. I know you would be a busy little woman for you would take a great pride in your wee house.

How is Fred progressing? Has he recovered from the measles?

Perchance I may go to Cairo for a few days leave before long, what can I get you? Anything you would particularly care for? I've got a pocketful of money.

Darling, I am happy and well and sincerely hope you are,
Fondest Love,
Your Man

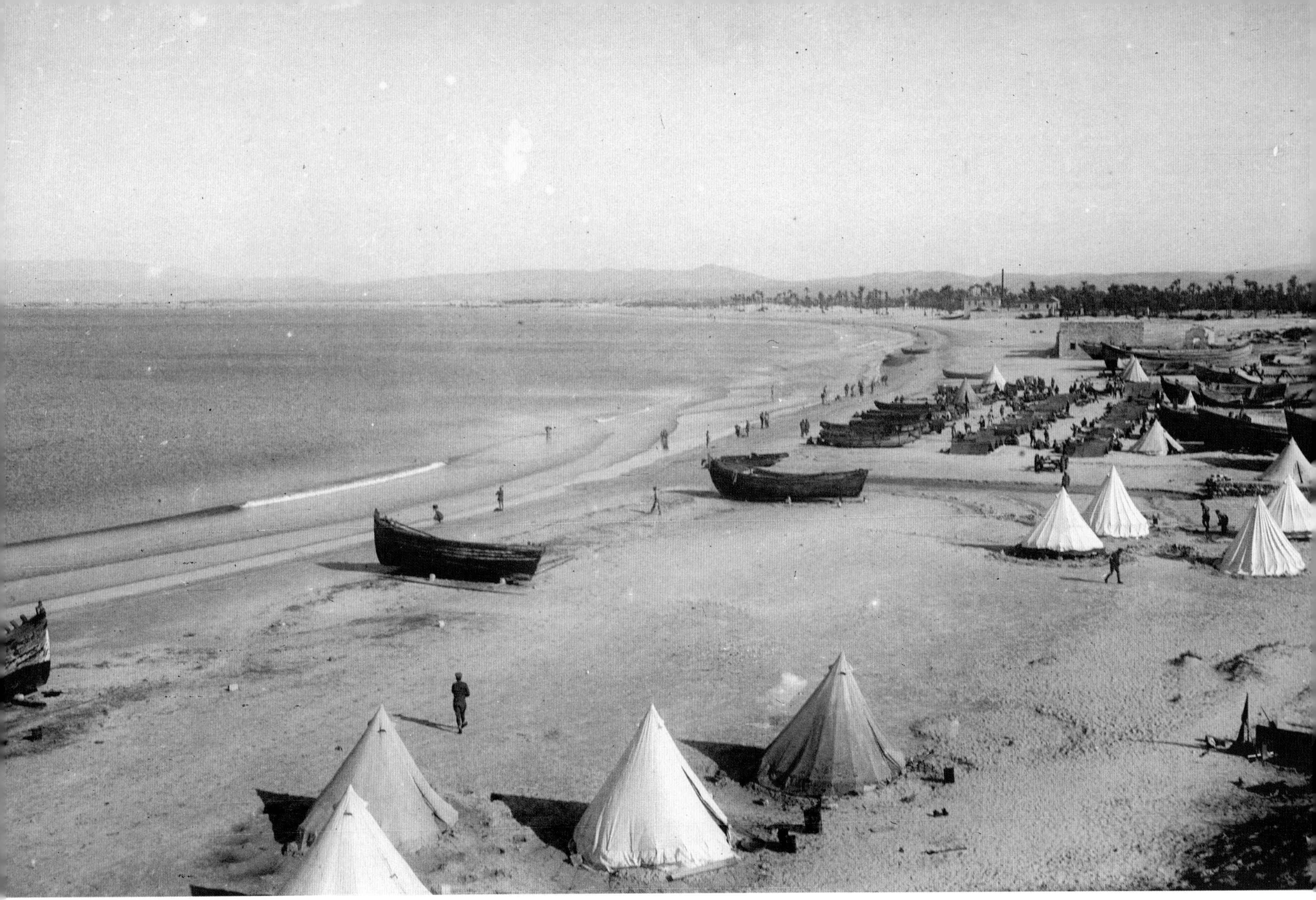

Tuesday 21.10.1917

Little Wife – Dearest

What an age it seems since I received you last letter. It's the worst of going in hospital – your letters follow me round from place to place and take a long time to catch up. I expect there will be quite a bunch of them – at least I hope so.

At present I'm at Convalescent Camp at Mustapha – a seaside suburb of Alexandria and very nice it is to get a few days' respite from the desert and duty. For 19 months I saw nothing but desert and duty and it's a great relief to get away from it for a day or so. I do not expect to stay here long! I expect they will bundle me off in 2 or 3 day's time and my one hope then will be leave in England. As I sit here on the seashore and write I can see white sails bobbing up and down on the water, and far away on the skyline is a steamer outward bound, and perhaps for home. Just picture me looking at it with a very far away look in my eyes and you can guess my thought.

I keep wondering how you are getting on. Are you better? I do hope so. I am quite well again and fit as a fiddle. It's a treat to be near canteens and shops where one can buy food and fruit for up Palestine way one can get next to nothing beside the army ration. So whilst here I am feeding myself and fattening my thinness – apples, grapes coconuts, pears – hot meat pies (in ½ hr's time), drinks, ices, Oh, it's a treat I've not had for nearly 2 long long weary years! It's really appalling what I'm stowing away – nearly as much as a camel – you wouldn't like to feed me for fourpence a day.

All this about my unworthy self and nothing about you – the dearest and sweetest of all. How goes it with you, my wife – mine, are you quite happy? And if not why not? How goes the hospital? Any business yet? And how goes Brightlingsea – do write me all the news. Anyone killed wounded or missing, and has Fred been home for his leave yet? And how is Moverons and the Lodge – I sent young Tom a PPC from Rafa a few weeks ago.

October – I suppose the leaves are fading and falling. How I should love to watch them flutter down with a golden sun sinking and shining upon the red gold and brown. But I suppose the trees in Victoria Place have lost their loveliness. Trees never seem to sleep here – leaves never fade and fall. No autumn, for the summer is eternal and the leaves and flowers always green and blooming, and corn grows and ripens in winter and summer alike. I should love to walk with you one of these evenings through the fields – say to Moverons and see all those glorious things – which for three long long autumns I have not seen. Shall we I wonder so next autumn, and when having had a lovely walk shall we wander home together to the house. Surely nothing can stand in our way.

Please give my love to your mother and also to Auntie Liza and please send me your love and a long letter by the next post

With all my Love,
I remain
Yours Always

"Off for a swim." At Alexandria.

10.11.17

My Dearest

Have received letters from you which have kept me alive during the past few days — for I've had and been having a tough time. I expect you've heard of the recent operations out here. I sincerely hope we shall be through with the job very shortly that I may have that opportunity to write you a letter.

Meanwhile I can only thank you for letters I have received, for they alone have kept my spirits from breaking. Dearest, I will write you a proper letter at earliest possible opportunity.

Ever Yours

[Undated]

... Often my desire for you is so great that it burns as a fierce white hot fire and I feel that I must cast aside everything and come to you, but it's a physical impossibility and I must wait, but I shall never rest contented until you are my little wife and we are together in our home. Today more than ever I want you and curse everything and anything keeping us apart.

This is a poor letter dear, I hope you will forgive such a wail but I feel a caged lion today and all I can do is to roar and rage and stride fiercely up and down my cage.

I hope you are in perfect health and spirits little wife and are as sweet as ever and that you are not losing patience in waiting for me.

All my Love,
And Always Yours
Your Man

16.11.17

What a war! It must be over a fortnight since I sent you anything like a letter. No dear, it's not that I have ceased to love you, or that I love you less, nor that I have fallen in love with a fair Syrian of Beersheba. On the eve of the Palestine push about a fortnight ago I wrote you. Since then we have had hard and rough experiences — unshaven and unwashed for days, and hard work day and night with little rest and sleep. Food is difficult to get, a dry stale dusty dirty hard crust of bread has been a luxury. Water is scarce and the sun has shone down pitilessly, scorching hot and choking blinding dust filling the air by night as well as day. It's been a purgatory. But our labours are rewarded, for Beersheba and Gaza have fallen and the terrible Turk is still running — may he never stop till he gets to Constantinople. Jaffa and Bethlehem and Jerusalem are in sight, and quite near to the tent in which I now write some Turks lie low side by side with feet towards the dawn, pieces of wood from the boxes of bully beef bear their names and numbers. There are Mohammeds, Ahmeds, Alhmuds, Husseins, Youssefs.

Often during these days and nights of grim struggle and desperate effort I have paused for a moment and thought of you. How I have wanted to write you, but only for a moment was it my pleasure to pause. Those thoughts of you have cheered and strengthened me and sent me on my way at times when I would have stopped and given way to despair. Never do I remember such strenuous times, never has fate been so unkind, yet I still smile and it's to you I owe these smiles. Some day, or winter's evening by the fireside, I may tell to you the tale of these days, then you will know what your love has been to me.

It's probable that you have been much longer than a fortnight with no letter from me, for I am told that just before the stunt all outward mails were held up and that previous to that a very strict censorship was on. Thus letters may have been destroyed altogether or at any rate held up, and you my dearest may have been a long long time with no letter. I shouldn't like to be so long with no letter from you.

For the time being all leave is stopped, but when this advance is over I rather expect we shall get some, and it's pretty well sure those who have been out so long will get leave to England. So I am hopeful, very hopeful. And you and I, little pal, we are going to marry! Make up your mind on that — that's dead certain, so just pull your little self together and accustom yourself to the name of Foster.

Now my pockets are fat with letters from you and I am richer far than he whose pockets are fat with corn or paper cash, for your letters are to me wealth untold.

"Turkish prisoners captured at Beersheba in the recent operations with BWA escort en route to Cairo". Possibly on the 1.05m narrow gauge line.

"Turkish prisoners captured at Beersheba in Oct 1917 at El Imara Station."

"Turkish prisoners captured at Beersheba entrained for Egypt and asking for water."

"Note the medical man with red Crescent. Prisoners taken at Gaza."

"How about the little chap in the centre for a terrible Turk! Prisoners taken in recent fighting at Gaza."

The New Church did a lot for the service men stationed in B'sea. Socials, whist drives and dances. We had quite a good time, I got friendly with several of the boys as they came and went. They must have got bored with me, my conversation always about Tim out East.

The Australians had a very good orchestra. They often played at concerts, specially in the Church. Pisani played the harp beautifully. There was a good social and dance one New Year's Eve, at midnight we all went into the Church for a service with the orchestra. Afterwards Pisani (with harp) walked home with me and Mary and George, who for some reason were sleeping that night at Victoria Place. We had all enjoyed the harp so much we asked Pisani to come in to play a few more tunes. This was a mistake, the music disturbed father who was in bed. It was the first and only time that I can remember he was cross with me, shouting down the stairs "What's going on, stop that noise!" How dreadful we felt, we had to ask Pisani to go!

Most of the boys at these affairs were NCOs. Dr Dickin gave a private dance for some officers. One dashing handsome naval man Lieut Horsfield, all the girls were thrilled to dance with him. I sat out with him during one dance. I soon found out he was rather a dangerous customer, very charming, with a nice wife. I had no more dances with him!

In July 1917 I went to stay at Hadleigh with Tim's people, and was introduced to many Hadleigh folk. Tim's mother was very proud to show me off to her friends: "This is Timmy's young lady." Although proud, she was very sensitive about my youth, and would explain to her friends that I was older than I looked, my hair in two long fair pigtails.

While I was at Hadleigh Blundell cycled over to see me. He was always welcome at Hadleigh. We went for walks around Aldham & Kersey. This was when Alfred gave me my first lessons in philosophy! How I regret burning his letters to me and to Tim.

I then went on to stay with Jess and her two babies, Audrey and Lennie. Poor Jess was having a hard time without Len, who was in India.

One night enemy aircraft came over Ipswich. Jess and I and the babies spent the night under the dining room table. Next morning I went down to the town, phoned Tom at the shop to tell him we were alright. This was the first time I had used a phone, the number was 13.

Alfred Blundell. Etching, in khaki, with Pollie.

A day or so ago I sent you a brief note with a snap of some Turkish prisoners. Did they reach you safely?

Your first letter was written from Leondith and posted on 28th Sept. I fear you must have had a rather dull time at Leondith what with the babies, the homesick maid and the sadness Jess must have felt at Len's departure and no one to take you out and about. Am glad you wrote me so fully for when Jess and Mother write they say but little because they say they expect you have written and told me everything. The Leondith honeymoon is not off, for I shall take Jess and the oxo cubes and the homesick maid bag and baggage over to Hadleigh. It will suit them and it will suit us and in Leondith you and I will soar to giddy heights. You tell me of a wee talk with Jess — she certainly advises us to marry etc, thinks your M and F will object. My dear little pal no one has ever accomplished anything this way. Men reach their goals by going forward and overthrowing each and every obstacle blocking their way, never halting or resting until their object is gained. This determination, this spiritual force, I have, and so just square yourself for marriage when I get leave and do not allow any doubts to darken your soul.

The next letter in succession was written at Hadleigh and posted at that ancient town on 8th October in the year of war 1917. It's a long long letter and you are a dear little woman for writing it. You appear to have been more cheerful at Hadleigh but although you had a dull time at Ipswich Jess was very glad of your companionship at that time.

Oh, I should love to see you and Blundell together both as silly and silent! Fancy him in a top hat, he looks a rum card in ordinary hats but in a top hat he must fairly take the bun.

And so you saw Gertie. Didn't know you'd seen her before. Gertie too is a funny girl — as funny as Blundell. Blundell on his visits to Hadleigh used to act the fool for fun and poor Gertie used to get most embarrassed. Silly Gertie, giving you kisses for Blundell! I reckon they were more acceptable to him through you than had she given them direct, for it would be your lips that gave them sweetness and I guess old Blundell rather liked the flavour of the sweetness. But it's mine, all mine, he just sipped where I drink. I should not be surprised if you even thrilled him, for I have heard from him the song of praises of you and called me a lucky chap. He sang to you too? He's a plucky chap. I'm glad you like him and he likes you for he is one of the best. It's very kind of him to invite us to spend our honeymoon at his shanty. It would be a very rustic honeymoon, the simple life indeed, but I know we should have a real happy time. We shall have to pay him a visit from Leondith. What fun it will be, we will have a real good time. I'm sure married life will suit us. We will lift old Blundell up among the stars for a while and just give him a peep at happiness such as heaven above can give.

Next time Doll asks you if we are getting married when I get leave just tell her yes. It's evident the proposal would not take them by surprise eh? How long have we been courting now? Is it 4 or 5 years? We ought to have been married in January 1916. I'll bet you a gateleg table and a plain gold ring that when your mother gets my letter she will say yes. And so our pal Blundell has written to you and you would like to know what he writes to me. Just half a minute dear, I will just glance again at his last letter. Oh! I really cannot repeat his words (you've told me take care not to spoil you). He thinks you're a dear and sings quite a song of praise of you, and every word he says is true.

You quote Blundell's remarks re my leave and submarines. Darling, no submarine will ever sink me on my way to my beloved — my wife — and no fear or foe will ever stop me once get the chance to come to you.

You wonder what my mother said about you. My mother is a funny old soul but she's a good old sort. She likes you very much. Previously she has liked you but thought you very young, now you are grown into a sweet young woman she looks upon you as her son's future wife and thinks you the very girl for me.

No, I do not think you may meet me when I come home, we must meet alone, all on our own, for it will take us half an hour to recover from our first embraces and then we shall want another half an hour in which to compose ourselves.

These are rough times out here — the grub is hard tack, the biscuits are very hard and the bully is stewed in its tin by the heat of the sun. Dust flows all the day and smothers everything. Every mouthful is well sprinkled with dust and flies and there are all sorts of creeping and crawling creatures about — lizards, chameleons, scorpions, centipedes, beetles etc. Do write and tell me what you are going to feed me on at Leondith on the honeymoon. Perchance, closing my eyes here, I may fancy the table spread before with good things.

Thanks for sending Auntie's love — please give her mine and tell her I shall be home some day and want a cup of tea after dinner.

It's very interesting to read of the dance and what a success you were. Oh, I should love to have eat in an obscure corner and seen you waltzing round. And you give all the credit to the dress! Just like you! I'm very glad you went and that you had such a delightful time. I want you to keep young, so dance and sing and be glad. Our love must not fix limits to your gladness but rather make it infinitely great. I do so want you to have all the happiness heaven can give you and I want to be with you always and to devote myself to you and your happiness, away from you I feel restless and uneasy and am always wondering is it well with you. Of course I know no one could be better loved and cared for than you.

    This brings me to the end of your letters and lovely letters they are too, you my love, my little wife. You are a darling and I love you with all my heart and soul. I long for the time when you and I shall be together never again to part. How glorious that will be my beloved, we shall love, comfort and take care of each other and share all our joys and sorrows, and when tired and weary you will sleep safe sweetly and soundly in my arms.

    Darling
    I am All Yours
    Yours Alone
    And Always
    Your Man

"By the well and seashore near El Arish. 30.9.17"

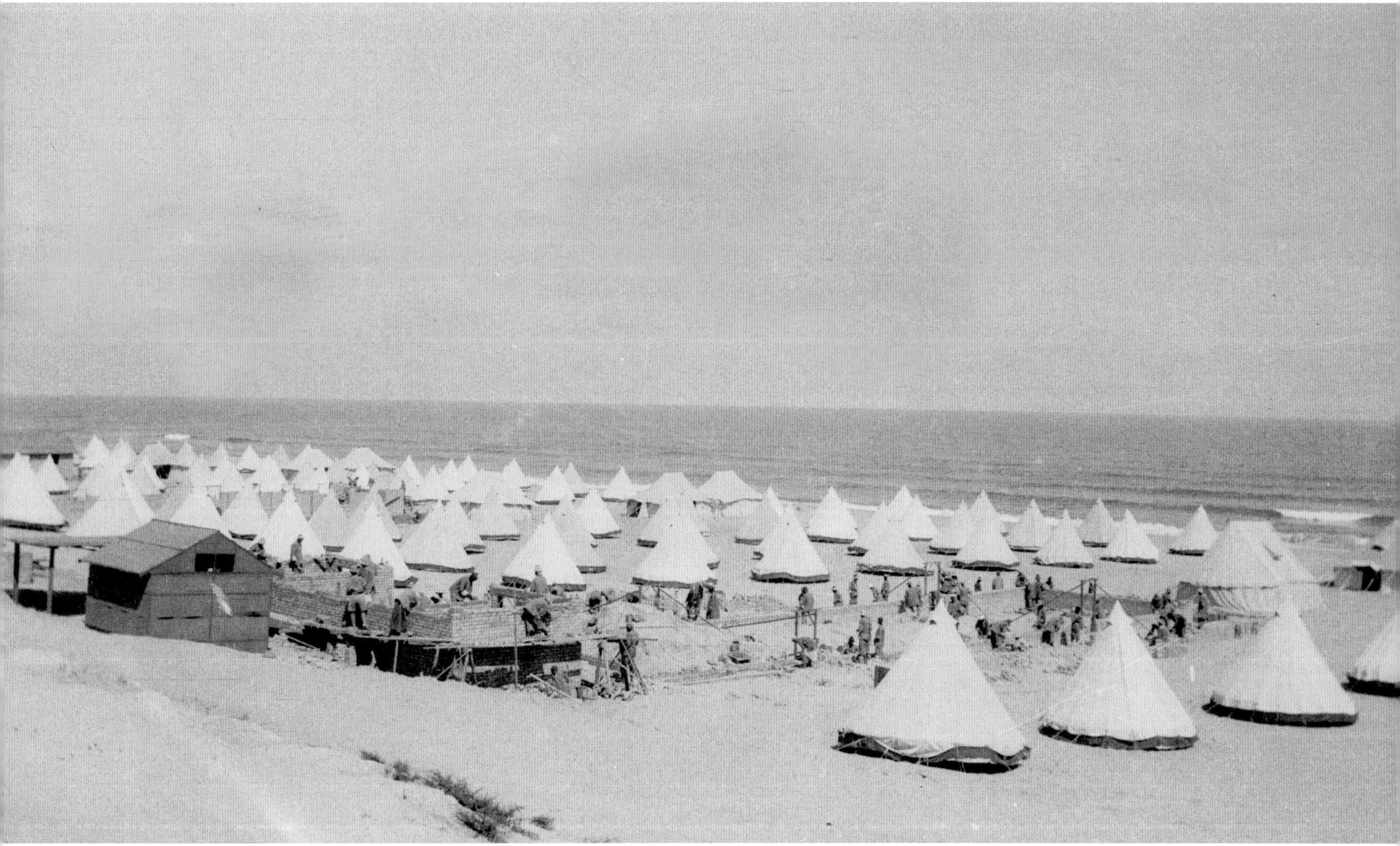

Special Christmas Number. Price Twopence net.
Part I — Double Number, Double Price
Saturday 1.12.1917

Wife Mine

I feel like laughing for pure joy — somehow or other I see the funny side of things when others so often see trouble. I'm in a particularly good humour, I'm extra especially happy. In similar circs some poor fool might have said "damn", but I — from an unfathomable place within the heart of my heart — I breathed a blessing. And everything within and consequently everything arising from that unfathomable place is yours and for you alone. Now you're puzzling your poor head trying to unravel this parable, for sure a parable it is. For is not a parable an earthly description for something heavenly?

My dear dear little pal, let me start this letter again. At such a tremendous height was I flying that my mind got giddy. Hence the flight of words, even now the tendency to shoot up into the heavens is so great I find it difficult to write as I would. My heart seems to take complete possession, right from the tip of my toes to the top of my head. My heart seems to be beating and singing with joy.

Probably you've heard of the Nile Barrage at Aswan near Cairo (one of the wonders of the world). Its flood gates hold up a whole season's supply of water for Egypt, and at a certain time those flood gates are opened, and thundery and cracking roar the waters, the life blood of a nation. For a whole season those waters have rolled from far equatorial regions and deep from the very heart of Abyssinia, 1000 miles or is it 2000 miles away. The whole year round from the canals and dykes, the land and people are watered. The whole year round that great barrage holds and stores its flood, until again those gates are opened and the swelling roaring flood crashes down and inundates the land, again filling with another year's supply the canals and dykes. It's the life blood of a whole nation and even we here in Palestine drink of the very same flood.

Now dear, what a gigantic thing! Is it not wonderful? Yet no more wonderful than my love for you — wife of mine. In fact of the two, the Nile and my love, I know my love to be the greater. And one of these days I shall open my floodgates and you will be simply inundated. Far away from equatorial regions rolls a flood, a flood of which even now you drink, and even now you thirst for more — thirsty girl you are. But one day those floodgates will be opened, then around and within you will be such a thundering of crashing — and my darling will be swept quite off her feet and carried away on that flood, and that flood will bear her safely, for that flood will be my love. And I laugh! I feel most amazingly bold! You have read of the gallant young knights of old — how they went forth laughing and bold — gay young sparks were they, and if some wretched old sinner stood in their way, they just poked their sword through the region of their heart and then with a smile and a song, stepped lightly along. They laughed and were bold and I laugh and am bold too. Why? Just because I love a certain little lady and she loves me and she writes all her doubts and fears — she doubts this, fears this and "this" when interpreted means her marriage to me and those doubts and those fears makes me feel like those warriors of old and I feel a great longing to sharpen my sword and tighten my belt. And of course I laugh and say, really and truly Poll, if I could get to you today — d'ye ken what I should do? I should fix it up with the parson at once, should threaten him with sudden death and bring him along to church. I would tell your people if they wanted to see us married they must hurry up for we were tired of waiting and were in a hurry and then off we would go, and by Jove! If they were not quick they would be too late. What should we care about mere conventionality? What do we care about trousseaux? We don't even care about wedding cake, not a little bit. We've sacrificed conventionality to our happiness every time and never regretted it.

Now what's all this about and what is it that stirred up such a stir within my heart? It was three letters. It was about 10am and my head was buried beneath a blanket and I was asleep (for I had been on duty all night) — very lightly asleep and dreaming of my sweet love — my little wife. Some fellows in similar circs would have said "damn" but I from an unfathomable place within my heart of hearts breathed a blessing.

Those letters are the most delightful and lovely and sweetest of letters and you — you're an angel, for you can soar high and therefore must have wings — spiritual wings — it's true but what are the wings of angels but spiritual wings? Now I've not really started writing this letter yet, I've merely been writing the preface — sort of letting off steam. Soon I shall find my feet and then will start. Now if I put the same energy into the letter I write your mother and throw my soul into it too don't you think I shall be successful? My dear I shall storm the position like the British stormed Gaza in recent days and as the British took the Heights of Abraham, and you will be mine as even Gaza and the heights of Abraham are now possessed by Britain.

All this is mere nothing to what I want to write. I've got yards and yards to write — a heart running over and a mind full of thoughts — but time, time, it will fly along. I feel as if I should like to catch hold of it, hold it up while you and I go past.

Dearest, this is to be continued if not concluded — continued — the very first chance I get.

Dearest
I am
Your Man
A Merry and Happy Xmas

16.12.17
Sunday Afternoon

Mine

The Sunday afternoon — just about the time when in the Sunday afternoons of years ago, I should be waiting and wondering if you would come and give me a look. Sometimes you came and sometimes you did not come. When you came — you and O. as a rule — we would walk up the line and I would make you late for tea. This afternoon is just such an afternoon as we used to get in August or July — bright, sunny, warm, a soft sweet breeze drifting lazily along and a white fleecy cloud waiting for a breeze to move it along. Happy happy days they were. This afternoon you will not come but instead a letter has come from you and that letter is just you — just sweet and beautiful.

I think the sympathy existing between us is extraordinary. On several occasions you have been writing at almost the same time as myself. Of just the very same sentiments and using very similar words as I have. In this particular letter you tell me how very happy you are you "danced ran and sang" for joy. About the same time I wrote you of my happiness and my words were "dance and sing" for joy. Then again you refer to the times when you would ask me if I still loved you and to tell you again, and you say in this letter I have just received that even when we are married and sitting by the fire in our bedroom you will want me to tell you yet again and again. Why dear, that's just what I wrote and said not so very many days ago! Don't you think this wonderful, and do you not think it is good omen.

It's glorious to know that you love me so and of course I like to be told again and again, and wife dear, you have just your whole heart in your letter, my darling.

Nearly four hundred letters I have written you, what a volume! I bet it's nearly as much as Emmie has had from Harry in all the length of their courtship. Do you mind keeping them all? It would interest me extremely to see some of my very earliest attempts at writing love letters.

Foolish foolish girl — you die before me?! First take my hand Poll. Come close, our bedroom fire burns brightly. Come closer and get nice and comfy. Watch with me the fire burning. We are young, we stir the fire and it burns brightly, our lips meet in warm tender loving kisses, we sit and watch, we laugh and we talk and we love, our eyes glow bright as the fire and we are so very very happy. But time passes, the fire is dying, you press closer to me and I hold you tightly. I feel your face pressed to mine. In my arms I take you and tuck you comfy in bed, still

Pollie's social engagements

in my arms. You and I, so close together, your dear heart so close to mine, we watch the firelight flicker and fading on the walls. And so we — you and I together — we sink to sleep, sweet sleep, the end of a perfect day. What could we have better? You, my darling going alone?! And leaving me alone in this bleak cold world? No dear one, we — you and I — will go together. Mollie will be a big girl then, in fact a woman, and she will have her own heap of love letters and her own man. She will be happy in the same way as we were when first we married, and you and I will go out together. We will never never part, and you would not want to go alone. No. You will come with me dear

So Pisani is his name — poor Pisani!! Is he single or married? Do you think he loves you? I should like an intro to him for he's got good taste. Can he play the fiddle?

Two dances in one week and you ask are you too gay? Can you remember how I used to reason with you when you didn't know what to do, when you were in doubt? I used to advise you to follow your inclination, No dear I do not think you too gay. I do not remember hearing of this Lt Gibbs before, it's good of him to find a place for you in his thoughts. I shall be interested to hear how these two dances went.

Chapter II

    I really think this must be your birthday letter for I fear that if I leave it later you may not get it by the time of your birthday and so I will now take the pleasure of wishing you a happy birthday and many many very very happy returns.

    Twenty! I think twenty is the right age for a girl to marry don't you? Really you are just entering your twenty-first year Poll, there should be no difficulty about our marriage, for you are no longer an infant but a woman.

    I don't suppose you have noticed the revision of Army pay announced recently in the papers. Under the new arrangement a wife's separation allowance is paid by the army with no deduction on that account from the husband's pay. So you see your pay of 12/6 per week would be all extra, none of it being deducted from my pay, and in addition to that I shall get -/3 per day more for my three years' service. Thus I should be getting 1/10/11 and you 12/6 per week. So if you were able to save only -/1 per day out of yours we should be better off married than we are single. That's good enough for a wedding isn't it? Then again, if things turn out as we hope, it would be nice if you could go and spend a little more time at Hadleigh, if you cared to. And should perchance Richard come, if you thought best you could go Hadleigh. Perhaps you would feel easier if my mother would take care of you. These are merely first thoughts and expressed only for your meditation, for I think it best to be free in our exchange of ideas and thoughts. It will help us to understand one another and see ahead.

    I should liked on this particular occasion to have written an extra specially nice letter but I fear this is a very poor attempt. I will try and do better next time dear. I fear I shall never write you as I should like. No, I shall never be satisfied until I can always be with you and be no end of trouble to you, breakfast, dinner and tea, and then supper. But supper won't be much trouble, we will have that on the hearth rug. We won't be an old married couple for a few years. For the first few years we will not be Mother and Father but courting and loving with all the privileges and advantages of husband and wife.

    Fondest Love
    Your Man

TIME TABLE – KANTARA

Stringline train graph with stations across the top: KANTARA EAST (0K), GILBAN (10K), [20K], PELUSIUM (30K), [40K ROMANI], MAHAMDIYA (50K), [40K ROMANI], ER RABAH (60K), KHIRBA (70K); vertical axis HOURS 0–5.

...RA MILITARY RLYS.—

TIME TABLE IN FORCE ON AND FROM 5TH FEBRUARY 1917.

DOWN TRAINS. Stations.		No.13.	No.27.	No.37.	UP TRAINS. Stations.		No.22.	No. 34	No.12.
KANTARA E.	dep.	0555	1255	1755	BURJ.	dep.	0250	1450	2150
	arr.	0607	1307	1807		arr.	0315	1515	2215
KILO 6½.	dep.	0610	1310	1810	KILO 159.	dep.	0320	1520	2220
	arr.	0625	1325	1825		arr.	0340	1540	2240
GILBAN.	dep.	0630	1330	1830	EL ARISH.	dep.	0420	1620	2320
	arr.	0655	1355	1855		arr.	0440	1640	2340
PELUSIUM.	dep.	0700	1400	1900	KILO 144.	dep.	0442	1642	2342
	arr.	0725	1425	1925		arr.	0450	1650	2350
ROMANI.	dep.	0800	1500	2000	BARDAWIL.	dep.	0455	1655	2355
	arr.	0820	1520	2020		arr.	0520	1720	0020
EL RABAH.	dep.	0825	1525	2025	MAADAN.	dep.	0525	1725	0025
	arr.	0850	1550	2050		arr.	0550	1750	0050
KHIRBA.	dep.	0855	1555	2055	MAZAR.	dep.	0625	1825	0125
	arr.	0920	1620	2120		arr.	0650	1850	0150
EL ABD.	dep.	0955	1655	2155	ABU TILUL.	dep.	0655	1855	0155
	arr.	1015	1715	2215		arr.	0720	1920	0220
SALMANA.	dep.	1020	1720	2220	SALMANA.	dep.	0725	1925	0225
	arr.	1045	1745	2245		arr.	0745	1945	0245
ABU TILUL.	dep.	1050	1750	2250	EL ABD.	dep.	0820	2020	0320
	arr.	1115	1815	2315		arr.	0845	2045	0345
MAZAR.	dep.	1150	1850	2350	KHIRBA.	dep.	0850	2050	0350
	arr.	1215	1915	0015		arr.	0915	2115	0415
MAADAN.	dep.	1220	1920	0020	EL RABAH.	dep.	0920	2120	0420
	arr.	1245	1945	0045		arr.	0940	2140	0440
BARDAWIL.	dep.	1250	1950	0050	ROMANI.	dep.	1025	2225	0525
	arr.	1258	1958	0058		arr.	1050	2250	0550
KILO 144.	dep.	1300	2000	0100	PELUSIUM.	dep.	1055	2255	0555
	arr.	1320	2020	0120		arr.	1120	2320	0620
EL ARISH.	dep.	1400	2100	0200	GILBAN.	dep.	1125	2325	0625
	arr.	1420	2120	0220		arr.	1140	2340	0640
KILO 159.	dep.	1425	2125	0225	KILO 6½.	dep.	1143	2343	0643
BURJ.	arr.	1450	2150	0250	KANTARA.E.	arr.	1155	2355	0655

Numbers 27 and 37 trains will each be provided with two passenger coaches, and No.13 train with one coach.

Numbers 22 and 12 trains will each be provided with two passenger coaches, and No.46 with one passenger coach.

Map of the region from Gaza to Jerusalem

- Jaffa
- Wadi Deir Ballut
- Sinjil
- Nebi Saleh
- Wadi el Jib
- Ayun Kara
- Surafend
- Ludd
- Nahr Rubin
- Nalin
- KEFR MALIK
- 2 Miles
- Ramleh
- Shilta
- Deir Ibzia
- Beitin
- El Kubeibe
- Ram Allah
- Bireh
- Yebnah
- Zernukhah
- TEL JIZAR
- Beit Ur el Tahta
- Ain Arik
- Beitunia
- El Mughar
- Abu Shusheh
- Vale of Ajalon
- Beit Likia
- Beit Ur el Foka
- Nahr Sukhereir
- Beshshit
- Katrah
- Latron
- Biddu
- Nebi Samwil
- Jisr Esdud
- Burka
- Kuryet el Enab
- TEL EL FUL
- Esdud
- Et Tine
- Bab el Wady
- Saris
- Mt OF OLIVES
- Beit Duras
- Mesmiyeh
- Kustineh
- JUNCTION STATION
- Wadi Surar
- JERUSALEM
- TEL ET TURMUS
- TEL ES SAFI
- Artuf
- Bittir
- MEDITERRANEAN SEA
- Bethlehem
- El Mejdel
- Askelon
- Burberah
- Faluje
- Beit Jibrin
- Wadi Hesi
- Deir Sineid
- Tel el Hesi
- Arak el Menshiye
- Hebron
- Gaza
- Beit Hanun
- Huj
- Jemmameh
- TEL EN NEJILE

30

1918: JERUSALEM

The EEF captured Gaza in November 1917, and extended the line to Deir Sineid. The line north of here was running on the Ottoman 3'6" gauge. The EEF engineers began converting the line to the standard British gauge of 4'8½", but utitlised the narrow gauge in the meantime. Junction Station (Wadi Surar) was reached in November and Jerusalem in December 1917. Ludd was won by February 1918.

The line east from Junction Station stopped halfway at Artuf because of the destruction of four iron bridges. These were reconstructed by the 266th Company of the Royal Engineers and Jerusalem was opened to traffic on 3'6" gauge on 27th January 1918. The British found two very worn out Turkish engines at Junction Station, and with 6 engines from Luxor and 3 from Sudan, were able to use the Turkish line to Jerusalem.

Conversion to standard gauge was put in hand on 22nd April. Jerusalem was reached with standard gauge on 9th June and opened to traffic on 15th June. In May construction of a 60cm line started in Jerusalem north to El Bira.

Tim arrived in Jerusalem at the end of January 1918 and became station master. The city was a welcome change from the heat and dryness of the desert, and he wrote expansively about its beauty and that of its surroundings, and would dearly have loved Pollie to come out to see it. He was there until October 1918.

Repair to one of the bridges en route to Jerusalem. With original bridges sabotaged by the retreating Turkish army, the Royal Engineers had to rebuild them using trestles in order to reach Jerusalem by rail.

Map showing the Turkish lines at the end of October 1917.
From *The Palestine Campaigns* by Lieut-General Sir Archibald P Wavell CMG MC.
London Constable and Co Ltd, 3rd ed. 1941.

Jerusalem 30.1.18

Little Wife

    The Jews have a wailing place here. If I come across it I too will wail — no mail, no mail, no mail! It's enough to break a man's heart, I am spiritually starved to death. The last letter I had from you dear was written on 12th December. Since that one was received, another letter of earlier date has reached me in a parcel — and for that parcel and in reply to that letter I wrote you some days ago my warmest thanks and greatest love. I do hope that letter will reach you safely for in it were enclosures for Auntie Liza, your mother and Mary, all thanks, and in your mother's letter I wrote of matrimonial hopes. You will let me know if the letter reaches you safely dear. This is a cruel cruel war — no letter from you or anyone else for nearly a month. I know it's not your fault Mrs Foster, it's the Kaiser — curse him — I'll make a hole in his heart with my bayonet if I get near him.

    For the past two days and two nights have I wandered down through the valleys and o'er the mountain tops and now am sitting in the station master's office at Jerusalem Station. This afternoon I took a first peep into Jerusalem and found it very fine. Some schoolgirls anxious to try their English spoke to me and got quite pally. They could speak Hebrew, German, English and Arabic and were the merest kids of 10 or 11 years of age.

    This morning I had the most interesting journey of my life — through the mountains to Jerusalem. It was a wonderful ride, great giants the mountains are, and on the very tip top of one I saw the tomb of Samson. He has a fine place high up in the sky, a mighty man he is at home among the mighty — on a mighty hill he is held high in the mighty sky and even the wind must often blow mighty high. On other hills I saw monasteries and villages and on the mountainsides and in the valleys were vineyards. It's a treat to hear church bells chiming and to be once again in civilisation. Oh! That I had some films, what photos I could take now. I must try and get some up from Egypt, for alas! Poor Jerusalem between British and Turk seems well nigh destitute.

    I've really got a lot of troubles and trials but they all seem lost — eclipsed by my one great woe of no mail. This is but a hasty note and I forgot that this was my only piece of paper, and then I hear (I wish it were <u>you</u>) someone calling me, but I will write you again the very first chance I have a long long love letter.

    All my Love
    Yours Always

Area of Solomon's Temple. Mosque of Omar. Mount Moriah from inside the city wall.

Through the hills of Judah to Jerusalem

At Bittir stn – the last stop on the railway up to Jerusalem

"At a wayside station on the way to Jerusalem. At Bittir station, LSWR loco."

"Towards Egypt just outside Jerusalem – in the hills of Judah". Third rail visible.

"Baldwin 3'6" loco."

JERUSALEM STN

"Jerusalem Railway station is by the side of Mount Zion, a few minutes walk from the Jaffa Gate on the Bethlehem Road. My billet is the room with shuttered windows on the extreme right. Just tap on the window and if I'm not out you will find me here."

Jerusalem Station. "Note the sign. The 'Jerusalem' in English was cemented up by the Turks during the war. The cement was removed from over the letters by myself sometime after the occupation by British."

The front part of the platform canopy (the uplift) was removed shortly after the standard track was operational in June 1918.

Tim in his room.

Looking towards the engine sheds from the roof of the station with the Turkish lines. 4-4-0 commandeered from Sudan in the centre. Behind it is a Luxor-Aswan loco outside the shed, and a captured tank engine in the sidings.

Looking in the opposite direction with the 3'6" track before the light railway goods yard was built (where the tents are). In the centre are two tenders from La Meuse engines.

Above and below: "3'6" Baldwin locos blown up by Turks when evacuating the city."

"Turkish loco accidentally started by native cleaner – ran away and engine finished up in turntable pit." Engine is a La Meuse 2-6-2.

Jerusalem 4.2.18

My last letter told of my arrival in Jerusalem. Since writing that letter I have seen a bit of the city and its historic surroundings.

Gethsemane is a quiet spot in a valley beneath the city wall and terraced for a little way up the Mount of Olives. The Mount is steep and all rocky. From its summit can be seen the little town of Bethlehem, the Dead Sea and River Jordan and, beyond, the Land of Moab. Perched up on the steep rocky side of a hill near by is Bethany looking very picturesque. In the Garden of Gethsemane I watched the sun sink behind the church of the Holy Sepulchre, occasional country folk riding on donkeys moved slowly past on the Jericho Rd. Right round the city wall I walked and every step was full of interest. The Golden Gate, Herod's Gate, the Gates of Zion, Damascus and Jaffa. It's a great wall. But poor old Jerusalem, blest not with milk or honey, the people begging for food were thankful for a hard army biscuit. The shops have little to sell but souvenirs of the Holy Land, mostly worthless articles at high prices, and no change can be had for there are but few coins in the city acceptable by the British, and the few articles there are of any worth are at famine prices.

My brother Fred was in Jerusalem on 48 hrs leave a few days ago and I did not know it until he had gone away. For a few hours we must have been within a few yards of each other for I arrived the day before he went away. It's disappointing we did not meet.

Sunday evening 9.2.18

This afternoon I actually heard an organ being played in a church close to the station. It's the first I have heard for over two years and to hear church bells and chimes again is quite nice and homelike.

What a long time since I heard from you! I'm looking looking and longing for letters. My move to Jerusalem will delay them extra. No matter how long your letters may be delayed I have no misgivings, like a gentle tap to a piece of fine china, such trust gives the ring of soundness to our love. You're a good girl sweetheart mine, all the love and trust I have for you has been created by you, not only have you created it but it's yours — my love and my life.

Poll dear, I hope you are happy and well
Sweetest kisses, Fondest Love
Always Your Pal

"Eastern wall and Golden Gate from Jericho Road. Jewish cemetery in foreground. Looking across valley of Jehosephat."

Golden Gate. "It was not the Golden Gate nor the camera, but the tomb which was used as a tripod that caused this great gate to incline. The tombs are terrible, I'll tell of them another time. Puzzle – find your sweetheart!"

"Mount of Olives from the Golden Gate. In valley below is the Garden of Gethsemane and the Jericho Rd. Above Gethsemane is the Church of Mary Magdelene."

"Refugees from Es Salt and other places about the Jordan and beyond. They are entrained at Jerusalem station for encampment at Wadi Sura. Photos taken from roof of the station. Left: Front portion of train. Right: Rear portion of train." On 3'6" gauge.

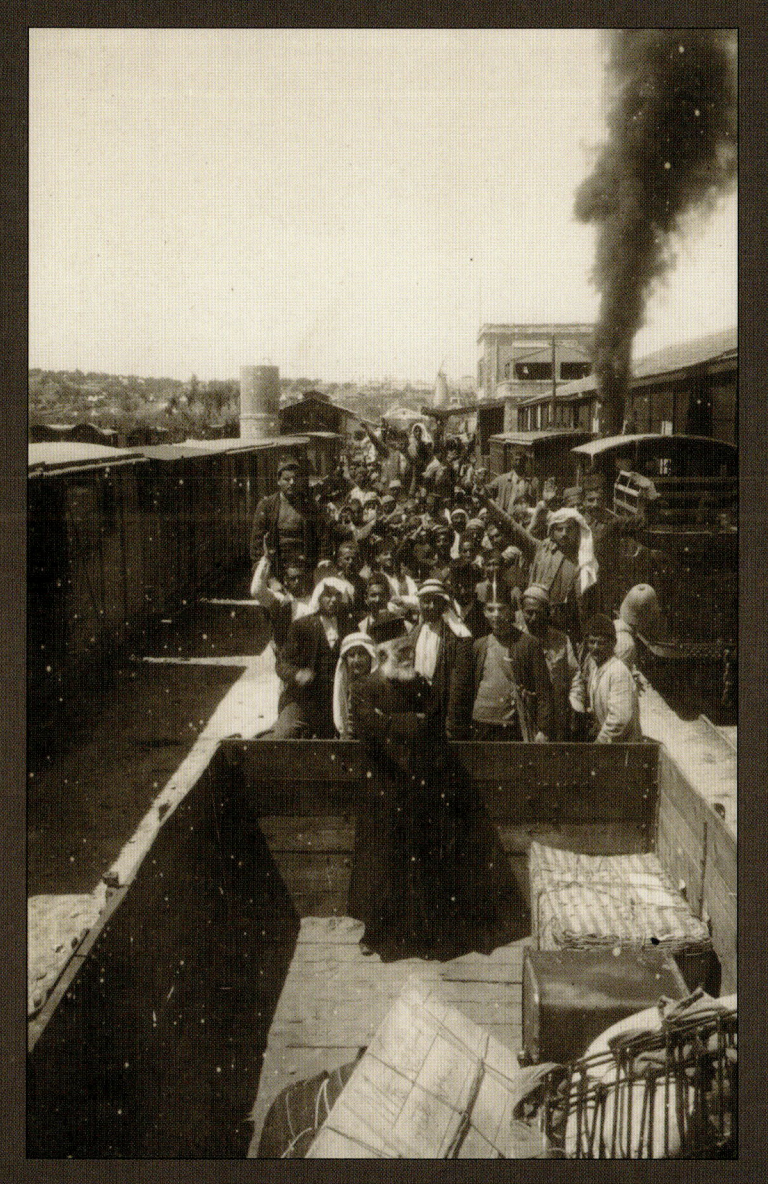

"Armenian volunteers for the French Army entrained at Jerusalem station for the base."

A later picture of recruits entraining to go south on the standard gauge track.

Sunday afternoon

I have told you of some of the sights I have seen here, they are a fraction of all I have seen. The other day I was following the way that Jesus went all the way from Gethsemane to Calvary and took photos of some of the scenes. In Gethsemane an old Franciscan friar gave me a little souvenir of that sacred spot and I am sending it to you with this letter. It's a little leaflet with the leaf of one of the wonderful old olive trees in that garden. I also took a photo of the garden with the old friar and a friend of mine.

The position of Pilate's house was pointed out to us and in the Church of Ecce Homo we were shown the original doorway from which Pilate addressed the crowd. Here it was that the Cross was taken up by Jesus and here commences that famous way of the cross — the Via Dolorosa. In the Church of Ecce Homo is a fine figure of Christ on the Cross crowned with thorns. Here also is a crown of gold and priceless jewels. Along the Via Dolorosa we were shown the successive stations of the cross and then Calvary — it's a most wonderful impressive sight.

Finally we went in the Church of the Holy Sepulchre. Here all Christendom is represented and a guard of Italian troops stand at the doorway. The decorations are magnificent and an hour or so is taken up in seeing right round the churches clustered together and forming the Church of the Holy Sepulchre. But most impressive of all is the actual Sepulchre itself. Only 5 can enter the small apartment at one time. No one speaks a word. Silently we stooped and passed through the little stone doorway. In silence we stood with bared heads and bowed, and in silence we passed out again each with a little lighted taper to see our way. Here too we saw the sepulchres of Nicodemus and Joseph of Aremathia and a piece of the pillar to which it is said Jesus was tied when scourged.

On another occasion I visited the Pool of Siloam between the valley of Kedron and that of Hinnom. Of this I have quite a good photo. The Pool of Bethesda I have also seen, excavations just previous to the war have revealed much of it that was hidden and still more remains to be unearthed. All these places are within a few minutes walk of the railway station.

"Garden of Gethsemane. Eastern wall of Holy City showing the Golden Gate."

Monday Evening
25.2.18

Time flies and this letter remains to be posted and you will be looking looking looking for letters. Although I fell very sleepy I really must finish this tonight and get it away.

Could you but have been with me this evening I would take you for a beautiful walk, for it's a glorious moonlit night. We would go along the Jericho Rd by the Virgin Mary's tomb through the garden of Gethsemane and back through Siloam.

I am sending you PPCs of the fifth station of the cross on the Via Dolorosa, Calvary, and the Holy Sepulchre.

Poll dear, I hope you are well and happy. I am very tired and sleepy and cold too and to bed I will go.

Good night Beloved
Ever Thine
Tim

"Mount Zion."

Mount Zion
Friday 26.4.18

Beloved

   On the sunny slope of the holy hill I sit and write to my best beloved. Away over the mountains I can hear the sound of great guns by the Jordan and before Nablus (Sheehem). Quite close to where I sit great David's tomb is raised, and nearer still is the room wherein "The Last Supper" took place. Many more historic buildings surround me. The day is warm and bright and the mountain air is fresh and invigorating and I — well — I would go clean off my head with delight to roam with you. It's great round here — great big mountains, deep valleys, steep deep ravines, great rocks — and for many miles we could roam over vast wild wastes of rock. Up on the mountain tops we could see for many miles — perhaps 50 or more — for we could see the Dead Sea and the river Jordan — Jericho and Bethlehem and Bethany too. I would take you in Gethsemane — it's a cool and quiet spot and sweet with flowers and other things. I know you would be very much impressed. Then we would scramble up the Mount of Olives and although you may be so wonderfully well and energetic you would want to rest on the way up — and you would find yourself not so good at climbing hills as I am. Wandering round and up and down the mountains has made me a mighty mountaineer! Then I would take you to see old Solomon, or rather his tomb in the courtyard of his old temple, and then nearby I would show you the village of Siloam where lived the wives of Solomon. Solomon loved the ladies and had a lot of wives and they all lived in a village and they were the sole population. How would you like that? You grudge the little love I would give to Richard and Mollie.
   This is my long weekend off. I finish at 2pm tomorrow till 10pm Sunday. Sunday morning I intend walking to Bethlehem. It's about 5 miles from here and you could come too. We would go to the Church of the Nativity, built on the site of the stable. They tell me Bethlehem is a pretty and pleasant place and I hope to get a decent photo or two there to send you. The camera I told you of sometime ago — the PC size one — is giving fairly good results.
   Now I've a great great grievance — since 1st Jan I've only had three green envelopes issued to me. Alas — poor me and you — you will not

"These are badly pressed and looked much better waving in the wind on the sunny slope of Mount Zion."

get this letter till I can get a green envelope to send it. Meanwhile I shall write other letters and send them in plain censored envelopes, so that you will get letters, but the letters, the letters in which I put all my love and all my heart and soul, these I will still write and send them all in a big bundle when I can get the green envelope. Oh, it will be a book! I think I shall have to get it printed and bound and call it the Epistle of Timothy and this will be Chapter I. But at the end of each chapter I should like to tell you how I love you, sort of a refrain — for I'm feeling very much in love with you, hence my desire to say a little song of praise to you. I think it's wonderful how well suited you and I are to each other — your sympathies are the same as mine. Had you been in Church with me last Sunday night you would have held my hand hard for the music was great — it was in the Church of the German Hospice on the Mount of Olives. It's a magnificent place and the organ is grand and glorious. How I should love to have had you with me — my heart's desire. You know how I love you dear. I've had two or three most glorious letters from you, they came last night, hence the lightness of my heart dear love.
    Here endeth Chapter one.

                                            Evening of the same day
    By the dim light of a candle I work now. The moon has risen bright and big above the mountains of Moab beyond the Dead Sea. I feel a wee bit sad and lonely. I feel that I want you to be here with me, I feel that you are the only one in all the world who could complete the perfection of the night, who could dispel the slight shade of sadness and make me perfectly and gloriously glad. Precisely you and no one else. I love you my little wife and I want you — a little child will weep and fret and no one but its mother can give it rest and peace. My heart is like a child — but in silence — for it cannot speak. But you — sympathetic soul — you know and understand, your sympathy, your wonderful understanding of me, your love is to me like a hand stretched out and clasped and so I am never alone. Your hand is always stretched out to me and Oh, what joy it is to me! How often when sad and lonely I clasp it still firmer and thank heaven for its blessing. Does it give you any happiness to know a little bit of what you are to me?
    I'm feeling particularly proud and pleased with you — if we were in our little room by the fire before bed I would whisper in your ear. I must close this chapter now dear.

Good Night my Beloved

Postcards of the Church of the Nativity

One of three Turkish La Meuse 0-10-0T engines built in Belgium, 3'6" gauge at Jerusalem, captured from the Turks.

Two La Meuse 0-10-0T engines. The inside of the cab is possibly one of these two engines.

"Jerusalem station before 3'6" track was taken up. Sudan government loco." The track in the foreground goes to storage dump.

3'6" gauge Turkish La Meuse 0-10-0T with unusual tender alongside one of the engine sheds. You can just see the platform of Jerusalem station on the left.

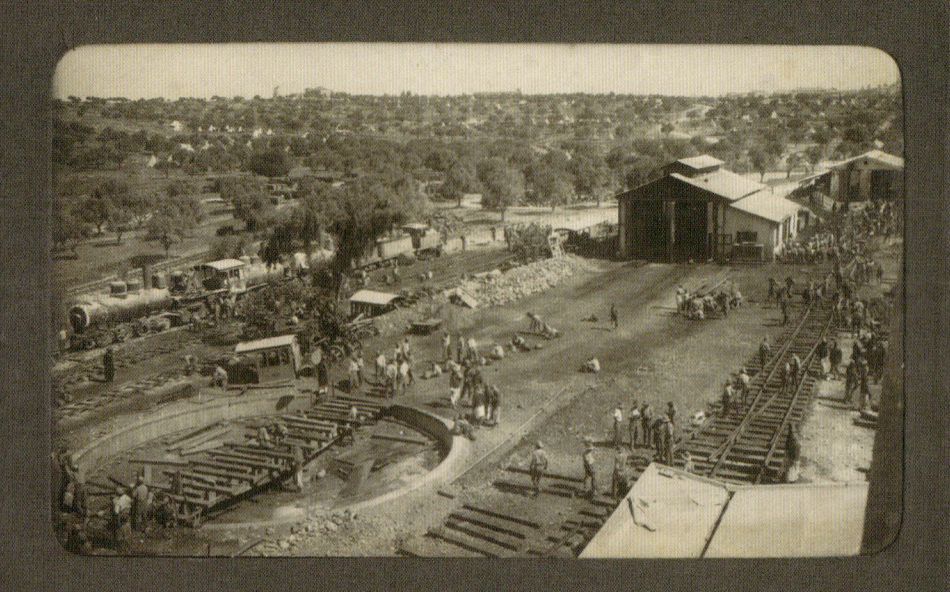

Standard gauge track being laid down in Jerusalem stn

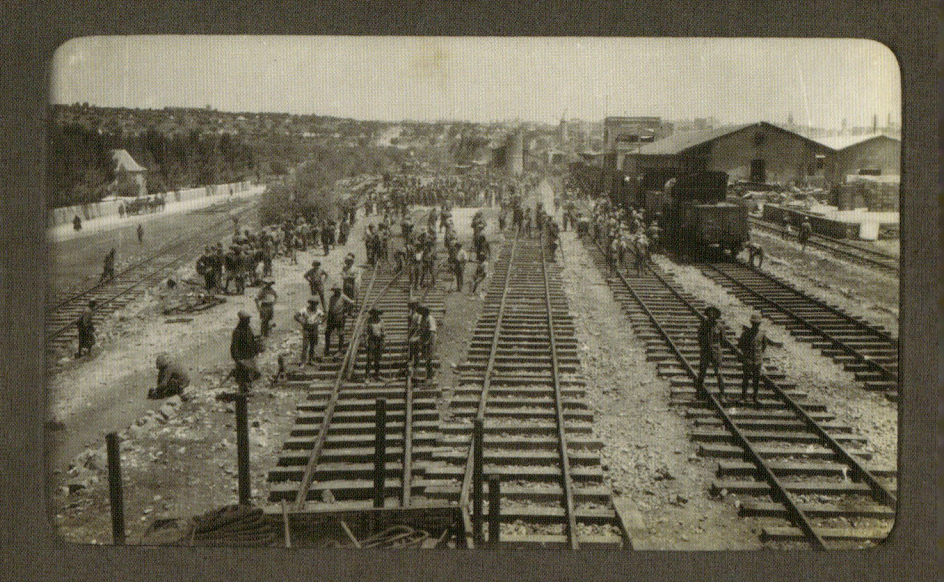

Standard gauge track being laid down in Jerusalem stn

"115 and 116 Railway Construction Company, RE and construction train in Jerusalem station – laying 4'8½" track in place of 3'6"."

"IW.+D. (Inland Waterways and Docks) Engines on construction train" which Tim was in charge of.

Standard gauge reaches Jerusalem June 1918
Tim in pith helmet having just made the first standard gauge journey from Cairo to Jerusalem (June 1918). Manning Wardle, No.27 commandeered from Inland Waterways & Docks. He brought sleepers and track for completion of standard gauge at Jerusalem. Tim's writing all over the tender.

"Shunting engine No. 99." Manning Wardle 0-6-0ST. This tank engine is thought to have been involved in a head on collision when it failed to hold the train and ran downhill towards Bittir and eventually collided at speed with LSWR 0-6-0 No. 444. The tank engine was practically demolished. One can speculate whether this was the accident Tim was a casualty in with a train carrying 700 Turkish prisoners. (See letter dated 11.10.1918)

The first 3 standard gauge lines complete. Front part of platform canopy still there, and "Jerusalem" sign in English still covered by concrete.

North

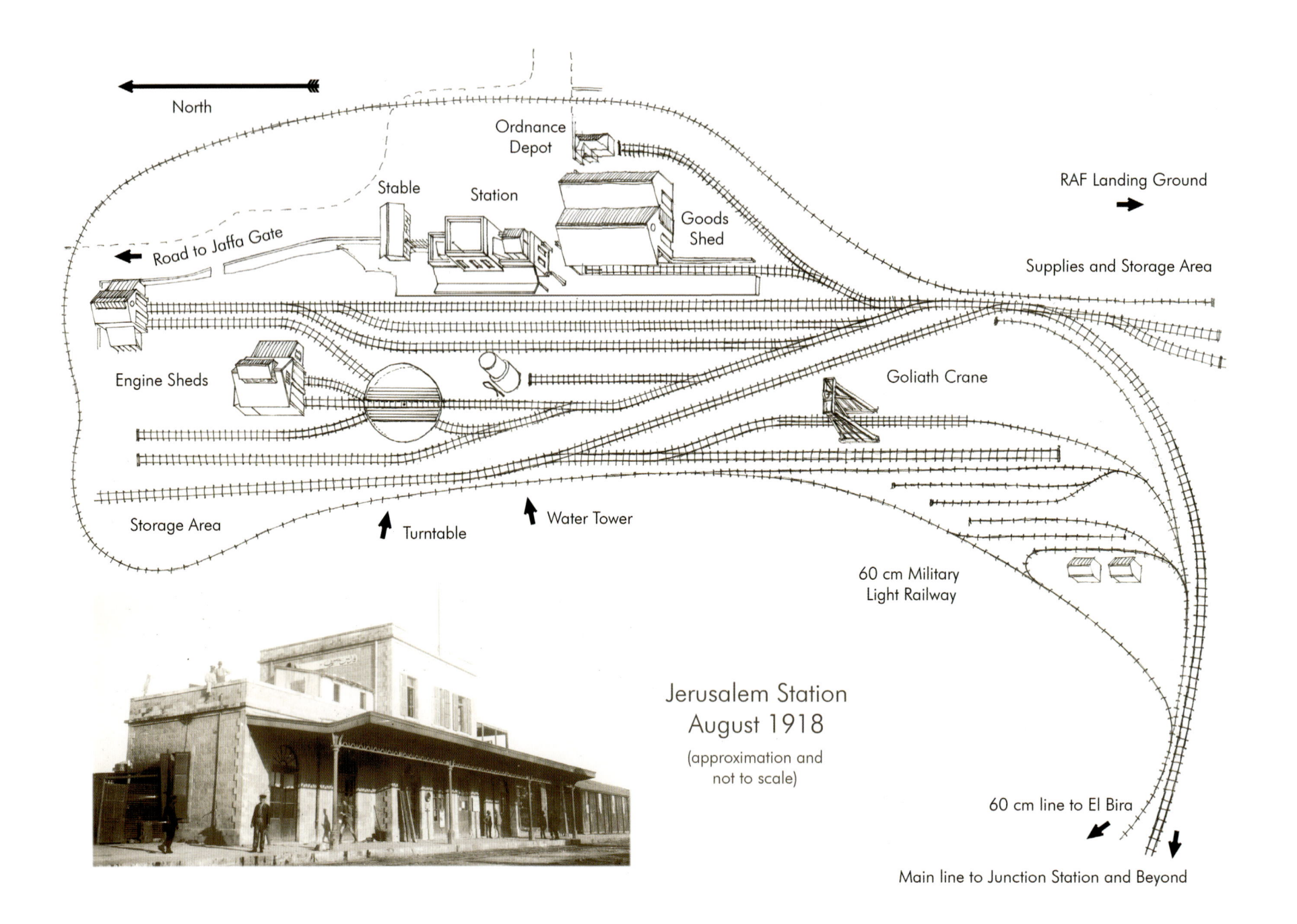

Ordnance Depot
Stable
Station
Goods Shed
Road to Jaffa Gate

RAF Landing Ground
Supplies and Storage Area

Engine Sheds
Goliath Crane

Storage Area
↑ Turntable   ↑ Water Tower

60 cm Military Light Railway

Jerusalem Station
August 1918
(approximation and not to scale)

60 cm line to El Bira
Main line to Junction Station and Beyond

The Light Railway

The 60cm gauge Light Railway Jerusalem to El Bira line was built by the Railway Builders Company 272 of the British royal Engineers, commanded by Colonel Jordan Bell, with some 850 Egyptian and local Arab labourers, about half of them women. It was started in May, completed 31st July and was 18km long. At its height in September 9 Baldwins and 7 Simplex engines were working on this line.

The light railway in Palestine was short of key skilled workers, especially guards and brakesmen. This was due to the sinking of the troop ship *Argon* off the coast of Alexandria on 30th December 1917. The loss included 76 men of the 96th Light Railway Operating Company, about half of whom were former railway men with at least 10 different railway companies. The very next day a further 74 men of the 98th Light Railway Operating Company were lost when another troop ship the *Osmanieh* struck a mine in the same area.

Baldwin 4-6-0T 603 on Jerusalem - El Bira line. One of 7 Baldwins that worked on this line in August. By September there were 9.

Looking down the track over the turntable with Goliath crane on right hand side.

"British Military Light Railway, 60cm gauge Jerusalem." 5 Baldwin engines and 3 Simplex 40hp petrol locos. In the far distance you can see a signal on the main line leaving Jerusalem where also the light railway leaves Jerusalem on a very tortuous route to El Bira and the front line at Ain Yabrud, although the last section was never completed.

Baldwin 4-6-0T 60cm gauge locos 611 and 587 offloading from standard gauge stock at Jerusalem station. Gantry "Goliath" by Ransome and Rapier Ipswich erected in July 1918. 2 Simplex locos in the distance. You can see the 60cm gauge running between the standard gauge tracks.

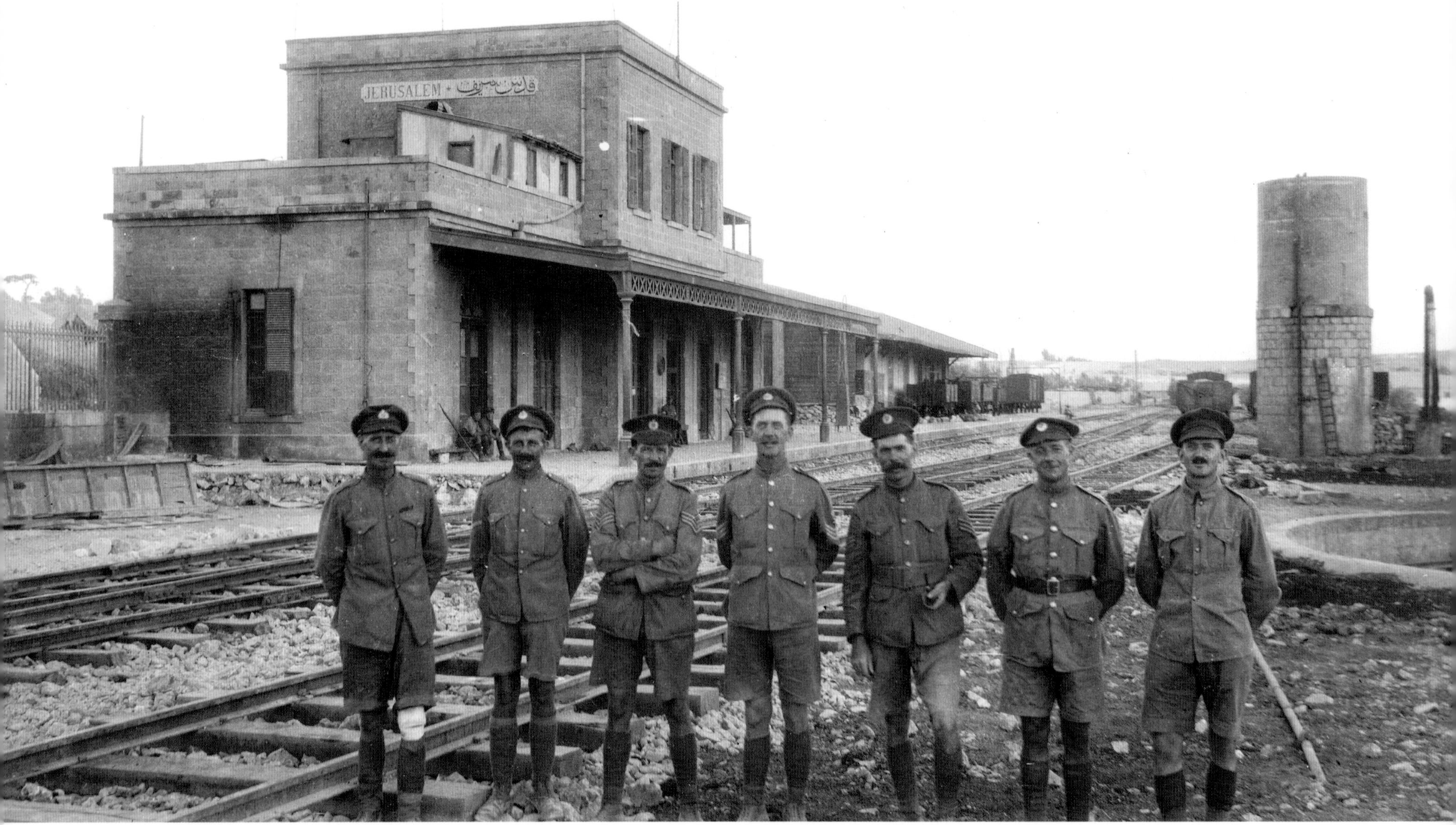
JERUSALEM
قدس شريف

Left: Staff at Jerusalem Station.
Above: "3'6" engines to work on turkish line Tul Keram-Haifa-Damascus."

Thursday 6.5.18

It's my turn to wash up and dirty dishes be all over the place. Without a doubt it's wrong of me to write when I should be washing up, but what do I care for right or wrong? And besides, that which is right from one point of view is wrong from another point of view, and although from one point of view it my be wrong of me to write when I should be washing up, yet from THE point of view I think the reverse to be true. And you ken what THE point of view is — beloved — there's only one point of view for us and that position is that there you are by my side, that's the position from which I view everything. I never see do or think without you, you my little Pal. You are so mixed up and interwoven in my mind my heart and my whole life that I cannot disassociate myself from you and I most certainly don't want to. Would you like to be here with me? Would you like to wander through olive groves, climb high hills and look down in deep valleys? I'd like you to see the orange groves, and the pomegranate blossom is beautiful. I wish I could send you a bunch fresh and unpressed — you'd like it. I wish you could be here and see with me all these things. Beautiful scenes, sounds or sights are made more beautiful by their association with those we love. A rose growing in the garden is glorious, but how much more glorious is a rose when worn by the one who is the only one in the wide world. You and I — we see suns set and think of the glory.

Jerusalem
Saturday 11.5.18

Dearest

There's one thing we can get in Palestine and get it good — that's oranges. I've just been up to the Bethlehem Road (that's a stone's throw from the station) to get some. I saw a dark damsel sitting by the roadside selling oranges. She wished me Neharak Si-ida (Good morning) and offered me two for one piastre. She was quite a nice girl and when I went, in addition to wishing me good day she added "Peace be upon you". That's a complement handed down from the days of David and is an expression of respect and affection in common use between Moslems, but rarely used between Moslem and Christian. So you see I can command respect and affection from Syrian Orange Sellers. These women are used badly by their men. But the oranges — they are delicious — big and weighty as cokernuts, sweet as sugar and juicy. I just eaten three — they are what is know in England as Jaffa orange.

I've just finished reading "The Light that Failed" by Rudyard Kipling. It's good, there are splashes and dashes particularly pleasing to me. They make me think of you and me and the muddy banks of creek and Colne. You and me playing the fool, wandering about all over the saltings and squelching over mud and water marshes, you angry with me for taking you down where water wetted your feet, and up the river too when we watched setting suns and blood red reflections on the water and listened to the plaplap of the tide on the shore, and away in the distance the murmur of the swell on the shore, and old Brown's boat riding up and down and swinging round on the tide. But these two, they did not marry. He loved and she didn't, and he was in Egypt and knew what it was to fight in the desert with mad dervishes and bedouins and he went to England again and he had a great misfortune and she passed out of the tale and he went to Egypt again and with a bullet in his head ended happily a hard life.

I was reading a piece in the paper pointing out that the purchasing power of £1 today equals not much more than 10/- before the war. That means £10 today is equal to about £5 before the war. That's a bad state of affairs. But in England girls — frivolous flappers — with no responsibility beyond their own dress, are being paid pounds for the most petty positions while Tommy — with a home and family — is paid pence for being butchered and bled with the glory of a cheap wooden cross and the shroud of an army blanket. To Thos Atkins belongs a great glory. He who fights in the line and gives his life for 6d or 1/- a day is a great fellow. I have tasted the horror and the hardship, but now I feel an outsider. I neither fight nor risk my life and yet get better paid than he who does. Not much honour or glory due to me and yet when I think how some civilians are getting fat and fatter still as the fighting goes on I feel that I would prefer to be with Tommy Atkins. And when the war is over those who have been fighting will be brought home to pay for the war and will be expected to keep the old country going and feed those who flourished and fattened. I've read in the papers of girls doing this and girls doing that as if these girls were making great sacrifices and bearing the burden of the fight. As a matter of fact what the girls are doing are getting very well paid.

There will be great changes in England after the war — greater even than the changes during the war. Thos Atkins released from the rigid discipline of the British Army will have something to say and something more to do.

Bethlehem

"Hospital train in foreground with a LSWR (London and South West Railway) engine goods train alongside it."
Beyond, 8 60cm light railway D-class bogie wagons arriving ready to be unloaded at the "Goliath" crane.

"The first standard gauge ambulance train to enter Jerusalem stn. Sick and wounded from Jordan valley being loaded." Goods shed and siding to the right.

"Motor Ambulances with wounded from Jericho at Jerusalem Station where ambulance train is being loaded for journey towards Egypt."

El Kuds Ess Sherif
Monday pm 13.5.18

Mine

That's the native name of the Holy City and if I were to write and tell anyone I was at EL KUDS ESS SHERIF they would never think I was at Jerusalem. It's a hot sunny day but there's a pleasant breeze stirring. I feel very drowsy but a desire to write you won't let me drowse. You have a photo of Jerusalem station — and you can see a sort of balcony — open air upstair arrangement — that's where I laze and write. Last night I went to church up Mt Olives and it was simply great. The organ thundered and roared in a devastating fury, crashed the whole earth into small pieces and then stood on the tip of its toe and trilled like a tin whistle or a nightingale. We had a short organ recital after the service, and the sun shining through the western windows blobbed the opposite wall with green and gold. Would you care to come with me some day? El Kuds is quite civilised. I wonder if you received that letter I sent with photos and flowers and a lock or two of hair. I hope you have for it was a longish letter. I've gathered some more flowers for you.

The mails seem to be very irregular. Today I have received your letter dated 5th April. It must be a fortnight ago that I received your letter dated 12th April and you do not seem to be getting my letters as you should, though I regret they have been not so many as they might had green envelopes been more plentiful. You tell me that you have paid 1/9 for a 4½ writing pad. Do you ken what it cost me to get some of my curls cut off in Jerusalem, <u>tenpence</u>! That's the price of a haircut and a very poor haircut at that. A boy earns ten pence for 2 minutes snipping at my hair. I really think I will let it grow in future and let it hang in long curls like the Jews here do. I am exceeding sorry to hear that Olive's Will is wounded, and sincerely hope it's no more than just enough to get him a trip to Blighty. You sing praises to my letters, and ask me how I manage to write them. It's wonderful what a fellow can do when he's in love — it's love that does it Poll.

You seem to be doing great things at drill. I've got a burning desire to come and cut you out and I shall certainly have to measure myself and if I can come across a weighing machine I will weigh myself and we shall see who is the tallest and the heftiest.

I am amused to hear of Old Joe and how for your benefit he sang songs of your "happy home". How is Joseph? Is he the same as ever. He's a rum old nut.

Saturday 18.5.18

Little Wife

How's B'sea looking. I feel quite an outsider now — quite a stranger. My only link with B'sea is you. Is the Portass pair still there? And the Australians too? And are all the girls married to the soldiers? I really believe there are more familiar faces for me in Jerusalem than in B'sea. How would you like living in Jerusalem? I'd be your pal and show you all the sights.

Is Doll still a bachelor and Dora too, and what's become of Ted Minter? And what's Jennie Bagler doing now and where does she live? How does the country look now? I suppose it's fresh and green, the roses and violets are at their sweetest and the songs of the birds too. Here in Palestine wild flowers are many varied and beautiful. The almond and orange grows too but the roads are dry and dusty, particularly with the heavy motor lorry traffic to and from the firing line.

I am
Yours Only
and Yours Ever
Tim

El Kuds Ess Sherif
28.5.1918

... Can you conceive any submarine sinking me? Oh, I almost laugh at the thought! I should never die so long as you live. And don't think you will ever care for the flowers on my grave, for when I go you come too. We — you and I — we're not parting, oh no! We go together. My sins are yours and yours mine. Your good works mine and mine yours. We rise and we fall together. Together we will see the sun set and together see it rise again. We take each other for better or worse, not because it is the law of the land but because we spontaneously desire to do so. "Oh that we two were a maying" is a beautiful song. We two are in the first verse yet we've got several more verses to get through and it takes years to get through a verse.

Other commandeered engines

"LSW (London and South West) engine at Jerusalem."

"ESR (Eastern and Southern Railway) engine at Jerusalem."

"Shunting engine No. 99 at Jerusalem."

"Captured by us in Sept 1918. 3'6" gauge wagon with aeroplane engine and propeller filled up by German railway troops. The propeller hauls the wagon along at about 25mph."

El Kuds Ess Sherif
Tuesday 4.6.18

My Love

Why is it that when writing to you I always want to sit out in the open air or else before an open window? Perhaps it is that out in the open air there is a sense of freedom and fullness, a sort of sentiment something to do with love. Oh! I've got it! It's infinity. Our love is infinite and to try and shut up something infinite inside something finite reduces it. Out in the open air I feel a freedom and fullness, and I am happiest when my love flows to you full and free — infinite. Do you know how I love you little wife? Just dream beloved — dream your wildest maddest and your happiest happy dreams, they cannot exceed the greatness of my love for you. Dream my beloved and you will get a glimpse of the love which I cannot put down on paper.

On the Church of the Holy Sepulchre is a stone said to be the center of the world. To the Christian world that no doubt is true, but to me, the centre of my world is in your heart.

Damn! Poll dear, you're a sympathetic little soul and I'm sure you'd say it too, for surely 1 or 2 letters from you must have gone down for ever. The following notice is published: "It is notified for information that a Mail from the UK and France for the EEF posted between the 25th April and 5th May has been lost at sea by enemy action". Sweetheart, I can stand the loss of many things without a murmur but heavens, your letters to me out here are just treasures! I simply live on them. Poll, now be a real little pal and just send me copies of all the letters you posted between those dates. You sweet angel.

I've a letter to answer, dated 17th April. You tell me of a letter... I was blue when I wrote it. Blue!! Poll. Oh! I'm badly blue sometimes — deep down in the very depths of despondency and my only want, my great longing, is for you. If I could only have you by my side, the deepest depths would have no darkness, no blues, for me. Yes wife mine I probably was blue when I wrote it. I endeavour to put in my letters my feelings as well as my thoughts. I think love letters should be so, I want you to see when you find I am blue, I want you to put your hand in mine and say never mind, and so you do. Oh, you're a little brick Poll! You're the sort of girl that keeps a fellow straight, that keeps a fellow on his feet when others would fall. You say you wish you had something interesting to tell me that your letters are all about yourself. Poll, you ought to know that it's about yourself I want to hear. So just tell me all about yourself — as long and as much as you like.

I had a letter from our friend George Ruffell. You remember the man of whom you were so frightened when you came to see me at the station. You remember when he thought you were a mouse. Oh dear oh dear! I didn't care a damn about my duty if you were anywhere about or even if there was the slightest possible chance of getting a glimpse of you. And old George never said a word, he's a decent dry old stick and writes me a very nice letter.

Later

I came across a PK camera today and gave 25/- for it and was offered 30/- for it an hour afterwards. But I'm no selling it, I've still got my PC camera but am not getting such good results as I should. I haven't got the proper hang of it yet. Out for a walk Sunday evening I was attracted by the beautiful blossom of a pomegranate tree and plucked for you a piece — all for you. No matter what I do or where I go it's all you — you're mine, all mine. Do you think of yourself as such and if so does the thought give you a feeling of captivity or freedom? I hope it gives you a feeling of freedom. I do not care to think that you may feel captive, I want you to feel free like a bird in the air but I want that freedom to bring you to me.

It's harvest time here. The corn, what there is of it, is golden and being gathered in. Apricots are ripe, oranges and lemons are ripening, pomegranates are in blossom, the vineyards are flourishing and yet sometimes my heart aches and I ask what's the whole world to me if I can't have you? Poll dear, do you love me as much as ever? Of course I know you do, but I'm like you, I like to be told again and again. I should just like to be by a blazing fire with you — just we two on a nice comfy settee or couch. Me first and then you in my arms, your head right high on my shoulder, and oh, how glorious it would be! Wouldn't we have a lot to talk about. And heavens, what glory to have my arm round you and your head so close to mine! Some might say we were soft silly lovers — what if we are? We're not going to turn sour if we can help it. You and I, we're going to just wander along and linger where we like and we like to linger and gather rosebuds so we're not hurrying along yet awhile. How I long to have your lips. Would you kiss me Poll mine? Could our lips meet now, would you put your arms round my neck? I have a lock of hair over which my lips often linger. If your head could but rest on my shoulder now my lips would linger on that hair.

Tim's PK camera
No.3 Folding Pocket Kodak
Model G
Film No. 118.

I'm glad dear that my letters give you happiness, that you find pleasure in reading them. Somehow or other when I write to you my heart seems to let itself go and simply pours itself out. What joy it is. It's glorious to know that my heart is yours and that you take my heart to hold it. I've printed one or two small slips with just my head on it, I thought perhaps you might give me a place near to your heart. At times when I feel worn out and weary I lie down and comfort myself with the fancy that I am lying with my head on your breast. Do you sometimes let me lie and rest in your arms? My darling, my beloved, will you let me lie and rest like that tonight? How I want you dear heart. How I long for you. Would you press your lips to mine? Could we but meet. Wife darling, I'm tired and not exactly up to the mark. I'm going to bed. Come to me tonight beloved come just comfort me...

                                              Jerusalem
                                              Midnight
                                        Sunday 30.6.1918

   Beloved

   The chimes on the church of the Holy Sepulchre have just struck 12. Little woman, I should like you to be here with me and hear them. I always want you to share experiences with me. I want my life to be yours and yours mine. Oh, when when will it be? Surely it can't be so long now dear. If it were only possible I feel I would come to you at once and never never again leave you. If you only lived in Jerusalem I feel I should just leave my post of duty right now and come to you even though you were sleeping. Oh, I would just wake you and tell you you were mine and that I would never leave you again. Would you be frightened my love or would you let me hold you in my arms and sleep again, kissed and caressed? Do you know how badly I want you dear? I think perhaps you do. I think perhaps you want me too just the same. How I love you!

   God bless you
   Little Wife Mine
   Your Man

"Jaffa Road: part of the western wall and the Tower of David."

"Jerusalem. Southern and eastern wall from Siloam Valley. Dome of Mosque of Omar in centre and Dome of Mosque El Aksa on left. Mount Moriah from outside the city wall."

Jerusalem
Thursday 18.7.1918

My Beloved — My Wife

How I long for you. More and more as time goes on I long and long for you. It would simply sicken you if I poured out all my longing on this paper. It would be like the bursting of a great big dam. Oh, my little wife! Just ask me if I love you. But you know sweetheart dear that I worship you. Within my heart, locked and sealed to anyone and everyone but my beloved one is all the love man ever had to give. Heaven has blessed me with a heart steadfast and true, and every beat of my heart is for you. Never once does it falter or waver but faithfully and unceasingly in true to you.

Poll dear, I love you so much I simply cannot write a decent sort of a letter, for my heart seems to fairly overbalance my brain and instead of my writing being regulated by my brain it's my heart that runs riot. My heart seems to say "Oh, give me the pen you cannot write for me!" And so my brain is thrust aside and my heart runs RIOT! And when my heart's let loose you know it's mad don't you? You know when we let our hearts loose — we both go mad. Oh yes! You're as bad as me and I as you.

Wouldn't I make a fuss of you if you were only here tonight. Oh, my beloved how I should love you! What a fuss I should make of you and I really do not think you would mind. No I think you would be quite happy.

The sun has left a glorious glow in the west and a soft sweet cool breeze floats in through my window. Little woman I just wish you were here with me. I should not be writing then, I would take my little pal for a glorious walk. In the hills we would go round great rugged rocks where no roads go, and high on the slope of a mountain we would rest. I would find my little wife a comfy little couch cut out of hard rock and soft with moss. Then I would wrap you up in my great coat and there you should lie and listen to the tinkling bells in the monastery near by. And if you liked it little wife, if you liked to lie and listen beneath the twinkling stars, when the bells ceased, then I would tell you all about my love for you and I sure by the time I had finished you would nearly be sleeping and if you could find room for me we would just get close together and we would sleep. My darling, is it not in my arms you should sleep every night? Every night dearest I want you in my arms. Once we are both in England nothing on earth will keep us apart.

Sunday 25.8.1918

Mine

I had a dream last night. You and I were sleeping, and when I woke up, flowing over my shoulder was a head of fair hair. My fancy wandered, I thought of the face pressed against my shoulder sleeping, I thought of how we loved, how right away in the early days we were drawn each to the other. Right away in the beginning our lives entwined like a vine — first a little tendril then fast they reached out, each holding the one of us to the other, and so we flourished and the vine itself entwined till each held up the other. That one could not be hurt or injured without the other suffering too. The life of each depended upon the life of the other, that if one died the other died too, the sun could not shine neither could the rain fall nor the frost bite but both felt the warmth, the sweet refreshing rain and the biting frost and the more the sun shone, and the more the rain fell and the more cruel the frost bit the closer those two vines clung together. And so in life and in death those vines were inseparable. I thought this to be an excellent arrangement of nature for neither desired to live alone. My lips caressed your hair which was flowing over my shoulder, my arm around you tightened and drew you still closer, gently you awoke and coming aware that I too was waking your head lifted up and our lips met in a long kiss and we neither of us wanted to sleep and so I told you of where my fancy had wandered and the talk drifted and you asked me to tell you of Richard and I told you of Mollie too. I thought R would be something like me — I wanted him to be of the best that was me and the best that was you. I would desire to see my little Poll again in M — just you and yet it would please me well to see something of myself in her — but mainly like you — just you again with a little bit of me. And when our day was dying, when the leaves were falling, when the fire was dying down and its light was flickering and fading, we — you and I — could sit by that fireside as we had sat when we married and we could reflect without regret, for had we not in both M and R so mingled our lives that in them you and I were mingled into one. And so we would watch the firelight flicker and fade away.

Little wife I've been a long way today. I've been to Bethany, to the house of Martha and Mary and to the Tomb of Lazarus.

"Recruits for Jewish Battalion entraining at Jerusalem station for equipment and training in Egypt. (July 1918)."

Monday Evening 26.8.1918

Dearest

I believe that which I wrote yesterday was a little inclined to be melancholy in tone. Yet I do not know why it should be. Such fancies should be light but in my clumsy hand a pencil is an awkward thing. In writing I am as one who grinds a hurdy gurdy compared with one who plays a beautiful organ. Still I will send you that which I wrote for I do not think you would have me destroy it. Oh! If I could only write like RL Stevenson, what letters you would get — why, I would write you a book!

Was it not in August that we were first engaged? But really Poll, we had no sudden dramatic engagement — spontaneously from deep down in our subconsciousness we understood — I did not go down on my knee and ask if you would be my wife. It was in our dreams we first conceived the idea. We dreamt — you and I — and the idea dawned gradually in our fancies — and we dreamed and we dreamed and I used to tell you to dream and dream and like the rising sun our dream grew brighter and clearer and gradually diffused our whole outlook — and yet scarcely a word had been spoken but we knew — you knew — and I knew, and I had no need to ask or you to reply for we each knew what the other knew. And when I put that ring upon your finger in the old first class, my heart was not in my mouth wondering if you would say me nay — for I knew. And our engagement and even our marriage is mere formality — merely the outward sign of and internal condition.

Little wife do write and tell me all about Hadleigh — how did you get there and how did you get away? Who did you see and how did you fare. I fear it must be a bit dull for you, there's not the entertaining society in Hadleigh there is in B'sea. I hope you got my letter re the ring alright and that you have got a ring that you like. I hope Jess saw to that. How's B'sea? Tell me all the news — write me a long letter. If at any time you should see either of the Portass's please give them my kindest regards

As ever all my love is always yours
Tim

27.8.18

Little Pal

What do you think? I've been playing TENNIS this evening, actually playing tennis! Why, I just went mad when I held in my hand a racquet! The court was concrete, and when smashing the balls about I was just away back in the days before the war in B'sea. My first "return", the first ball I properly got hold of I sent just about over the Mount of Olives and well on its way to Jericho. Gradually I got my hand in and very soon was my old self again — all over the place — turning somersaults, flying from one end of the court to the other. I had a single first, then a couple of Australians came and made a quartet. It was great, and we were on a high hill and played till the sun sank rosy red behind the high hills far away. And alas! We get but very little twilight here, nothing so long as in B'sea, and so we soon had to finish. And as I walked up through the dark streets of the old city I thought of those happy days when I used to change your slippers, putting on your shoes and stealing kisses — sweet kisses they were — and we used to slip off for a little walk or perhaps down to the old first class. I remember one time — for some time — you would not come and I could never understand why — it mystified me.

Is there any tennis in B'sea now? A woman from Australia has opened in Jerusalem a soldier's club — a recreation room, and there is a tennis court. Does Miss Ham play? And is she not married yet? How's Mrs Leater? And where's your partner — Mr French — now? And how's old Mr Hibbs and his 'taters and cabbages and lettuces, and what about Miss Purkiss — have you given her my kind regards, and where's Miss Fookes and Miss Whatshername who use to be at the Co Op with Miss P?

August 1918. Great news from Tim telling me that after 3 years in Palestine the men were to get 1 month's leave in England! His turn would be in March, what a long time to wait, but at last something for us both to look forward to. However, the war was to be over before then! Armistice was to be signed 11th November 1918. We thought then the men (those who were left) would soon be home, but it was many a long day before Tim returned, hopes were raised time and again.

On Armistice Day, the good news was to be read out from Colchester Town Hall. I and several other girls got a lift into Colchester on an army lorry and managed to join in the crowd outside the Town Hall. When the news came through everyone just went mad! We came back to B'sea on the army lorry again. That night there were great celebrations, everywhere was lit up, singing and dancing on the green (Victoria Place). One unfortunate incident: a soldier threw a hand grenade, which landed through the window of old Aunt Emma Eagles' bedroom, setting fire to the curtains and room, poor old lady paralysed in bed. This caused a panic, and a hush in the crowd. My mother was soon on the spot, the poor old lady lifted out of bed and taken to another room. It was thought that the shock would finish her. But she was to face a worse one some weeks later when her eldest son shot himself with his Home Guard rifle. Later her younger son died, the result of being gassed in the war and a heavy drinker afterwards. Such a nice boy in his youth.

Aunt Emma was a brave old lady, bed-ridden for many years, she had been a jolly old girl, never lost her sense of humour. My mother used to go along to sit with her for a while each morning and evening. I sometimes read a novel to her, a few chapters at a time. It had to be a nice love story!

Gen Sir R Wingate's special train, the first standard gauge train to run through from Cairo to Jerusalem, waiting at Jerusalem stn ready to start on return journey to Cairo
L.S.W.R. Loco, E.S.R. Coaches.

General Sir R Wingate, High Commissioner for Egypt inspecting guard of honour of Ex Turkish City of Jerusalem police at Jerusalem stn.

"The first standard gauge passenger train to run through from Cairo to Jerusalem station. General Sir Reginald Wingate, High Commissioner for Egypt. LSWR loco Egyptian state Railway coaches."

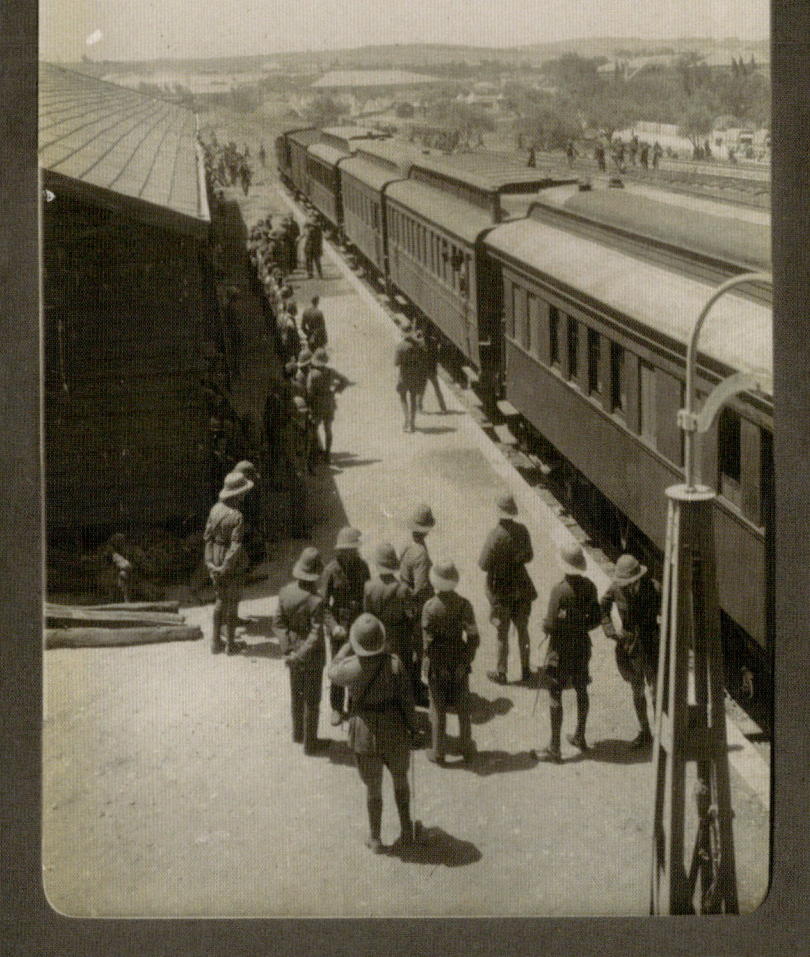

General Sir R Wingate after his visit to Jerusalem City about to entrain at Jerusalem stn for journey to Cairo

Turkish, Austrian and German prisoners waiting outside Jerusalem station for entrainment to Egypt (24.9.18).

"Refugees from east of the Jordan assembling at the station for entrainment to Wadi Surar on the *Shephelah*". Back of the goods sheds from the public square. Entrance to Army Ordnance Department.

"Turkish officers back in Jerusalem as prisoners of war. Waiting at Railway Station for train to Egypt (their land of bondage)."

Friday 4.10.1918

Dearest

It's a monstrosity – it's illogical – that you do not write me for THREE WEEKS! It's the refinement of cruelty – would you rather I died from starvation of the soul or German measles? Oh! I could fall asleep with a smile with German measles and a letter – but to have no letter? Why, I should die with no desire to rise again with a miserable expression upon my face. But bless you! [Must have just received a letter!] You did not deny me for the three weeks.

These are my last days in Jerusalem – I'm going up through Samaria to Galilee in about a week's time. Then for a time I fear my opportunities to write will be fewer than now. You will not get so many letters.

I've grown to like Jerusalem the Golden. Poor old Jerusalem, what sights it has seen. What joys, what sorrows, what weeping, what laughter. I've been here longer than many of the British. I've been here longer than any of the railwaymen. I've done more duty here than any, every day or night for nigh on nine months. I'm looked upon as an old timer in the Wild West. What walks I've had, what sights I've seen. There's some grand scenery in the kingdom of Judea.

Standing upon the summit of Olivet, one can picture the multitude listening to a sermon preached by Jesus. Away to the east some 30 or 40 miles one can see quite clearly the Mountains of Moab, where good old Moses stood and viewed the promised land with Jerusalem in its centre for its heart and soul. Then down, deep down, far below the level of the sea, flows Jordan, in between and due east of Jerusalem, being itself in the Bahr Lut or Dead Sea. Still standing in the same place one can face the south and see the little town where grand old David was born and spent his boyhood, where later on the wise men went guided by a star. South east and almost at the foot of the Mount is Bethany. Poor little Bethany seems nearly reduced to ashes and dust. But greatest of all, look towards the west: there is the Holy City spread before you like a panorama. There immediately before you is the Golden Gate, the Courtyard of Solomon's Temple, there is the Throne of David – great massive Halls and Walls. There on the left is Zion, there is the house of John and Mark wherein was held the last supper, and close by lies the Nebi Daoud, the tomb of David. A little bit to the right is Calvary, rising above it is a great big dome surmounted with a cross. Beneath it is the Holy Sepulchre. Winding its way up from the House of Pilate is a narrow stone paved street worn smooth and slippery with centuries of bare feet – it's the Via Dolorosa – the Way of Sorrow.

Immediately at the foot of the Mount of Olives in the valley between Mount Moriah and Olivet is Gethsemane, a sacred spot. In the same valley is the place where St Stephen was stoned and the chapel of the tomb of the Virgin. Follow this valley on its way south and we come to the village of Siloam. There on the side of the hill halfway up the steep slopes of Mount Zion, as if to rest and refresh those who struggle up those slopes is the Pool of Siloam, so cool and shady, clear as crystal and cold as ice, summer and winter. Throughout the long rainless sweltering summer the water flows here out of the solid rock.

This is a mere fraction of many aspects which go to make Jerusalem today, not modern buildings or sites, but the life. There never was such a mixture, I've been amongst it and have seen it. I've seen that which is gross and that which is gold. And yet to me all this is exile no matter whether it be in Jerusalem or Galilee.

Referring to my distaste for empty-headed swollen-headed swanking subs, you tell me there are officers who are "good sports", "one of the best" etc. Oh I know that right well! As surely and as well as I know of the other sort. Some I have seen stare right into the face of death and never turn a hair. Some I have seen go to their death with never thought for themselves. There are incidents I shall never forget. One I recall was in the dead of night... two days later he was lying on the battlefield dead. He never had a grave, not even the rough wooden crosses we used to make out of bully beef and biscuit boxes. He was an officer, no knotty swollen headed sub washer. I know others, some of them young subs, some of them old campaigners. It's not for me to attempt to glorify them, they are far above me. I can only look up and silently adore them. But the empty swollen-headed little snob! I'm lucky. I can so to speak laugh at them. They must not, they cannot command or order me or make demands of me. They may consult me, they may ask of me, they often ask favours of me. That is so far as my duties are concerned. Some in ignorance try their swank or bluff on me, and oh, does it not delight me!

I scratch and scrawl. Racing goes the time. Fast as my pen flies old time seems to go faster, and so far as my letters to you are concerned I can never get ahead of time.

This is but a small fraction of what I want to write to you this morning but alas I must conclude and continue another time.

Adieu – for just a little while
Fondest Love
Yours

"RAF landing ground."
BE2c mainly used for reconnaissance. In several pictures of Jerusalem station looking south down the line, aeroplanes can just be seen on the brow of the hill.

"Supply dump at Jerusalem."

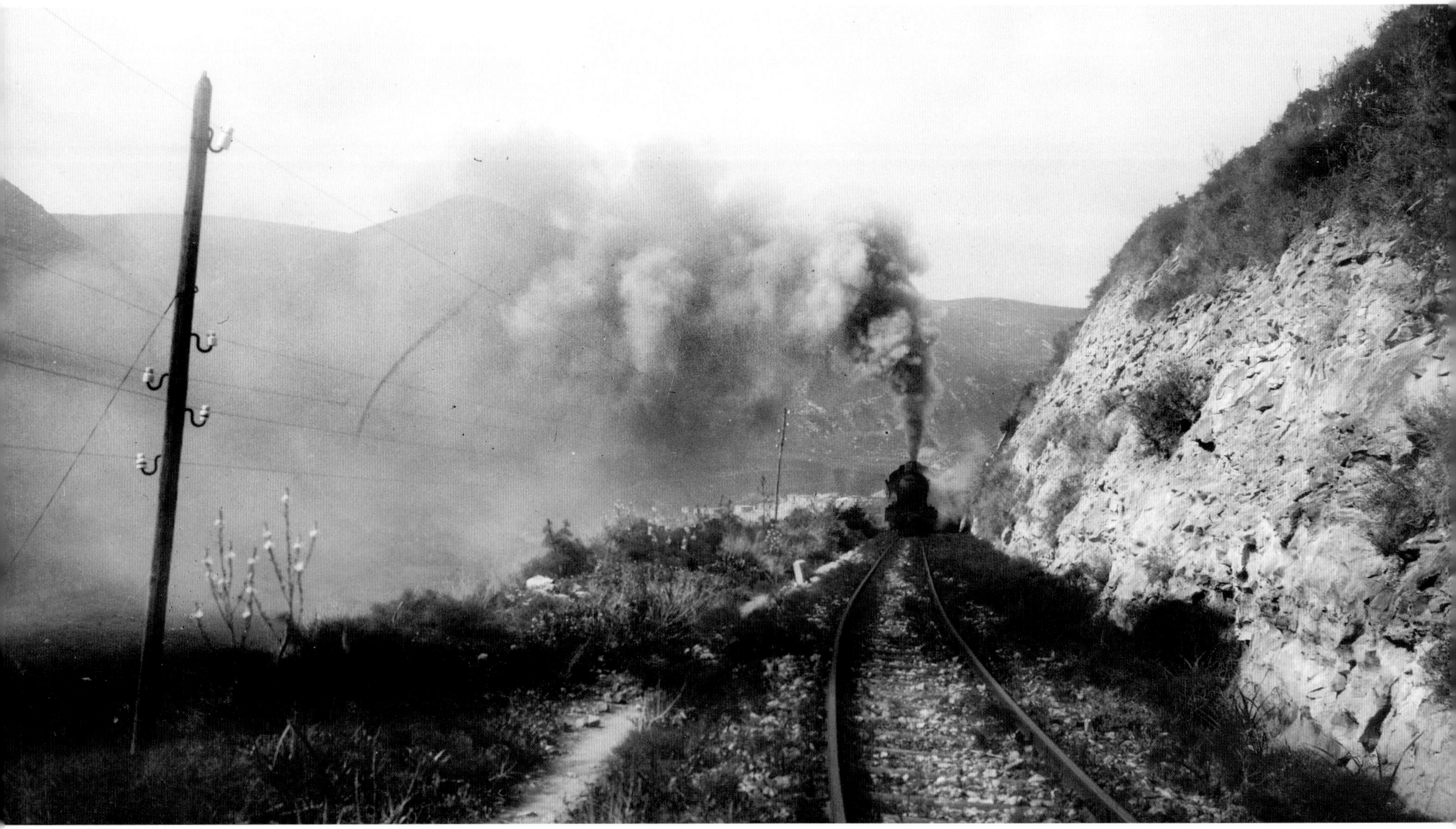

"The pre-war narrow gauge (3'6") Jaffa to Jerusalem railway near Jerusalem"

THE WEEKLY NEWSPAPER OF THE
EGYPTIAN EXPEDITIONARY FORCE
OF THE BRITISH ARMY IN OCCUPIED
ENEMY TERRITORY.

PUBLISHED EVERY THURSDAY
AT G.H.Q. FIRST ECHELON,
PALESTINE.
PRICE: ONE EGYPTIAN PIASTRE.

The Palestine News

FIRST YEAR No 39   THURSDAY, 21st NOVEMBER 1918.   ONE EGYPTIAN PIASTRE.

THE PALESTINE NEWS, THURSDAY, 21ST NOVEMBER, 1918.   7

THE ARMISTICE TERMS.

THE PREMIER'S SPEECH.

MILITARY CONDITIONS.

THE WESTERN FRONT.

...don, 11.— The House of Commons was crowded ...y part when Mr. Lloyd George rose to announce ... Armistice Terms. The Prime Minister, who was ...ved with a tremendous ovation on entering the ..., stated that the following are the terms of the ...stice:—

...ern Front operations, on land and in the air, to ... at eleven in the morning.

EVACUATIONS.

Swiss frontier. In the case of the inhabitants no person shall be prosecuted for having participated in any military measures prior to the signing of the armistice. No measure of a general, or an official character, shall be taken which would have as a consequence the depreciation of industrial establishments or the reduction of their personnel. In all territory evacuated by the enemy, there shall be no evacuation of inhabitants, no damage or harm shall be done to persons or to the property of the inhabitants.

STORES AND RAILROADS.

Military stores of food, munitions, and equipment are not to be removed during the periods fixed for evacuation, and shall be delivered intact. Stores of food of all kinds for the civil population, cattle, etc., shall be ...

provision. Freedom of access to and from the Baltic shall be given to the Allied Naval Forces. To secure this the Allies and the United States shall be empowered to occupy all German forts, batteries and defence works in the entrance to the Kattegat and Baltic, and sweep up mines, the position of which shall be notified.

ENEMY MERCHANT MEN.

All German Merchant-ships at sea to remain liable to capture. All Black Sea ports shall be evacuated by Germany, and any Russian war ships shall be released, and Allied merchant-ships shall be restored without reciprocity. No destruction of ships before restoration, and all Neutrals, particularly Norway, Sweden, Denmark and Holland, shall be informed, by Germany, that all trading restrictions with the Allies, imposed upon them by Germany, are removed. No transfer of German merchant-ships of any description to any Neutral flag. Armistice to last for thirty-six days with the option of an extension. During this period the failure to execute any clause of the Armistice shall render it liable to be denounced at forty-eight hours notice. (Loud cheers.)

After reading the terms of the Armistice the Premier said:—

"Thus comes to an end the most terrible and the most

1918-9: THE HEJAZ RAILWAY

We know from Tim's photographs that he was at Tul Keram soon after its capture in September 1918, and that he was station master there for a short while. The coastal Plains of Sharon lay on the low marshy land between Haifa and Jaffa and were rife with malaria, which Tim complains about in his first letter. Turkish Armistice was signed at the end of October 1918. After Tul Keram, Tim was station master first at Haifa and then Samakh at the southern point of Lake Tiberias (Sea of Galilee), on the Hejaz line which ran from Haifa to Damascus and Hejaz to Constantinople operating with Turkish gauge engines. Before these lines could operate the Canadian Royal Engineers had to rebuild the bridges we had blown up to cut off the Turks' retreat.

In total, 627 miles of standard gauge track had been laid by the Royal Engineers, with 748 points and crossings, and 86 stations built. At its peak, 169 locos operated on the Palistine Military Railways, and the ROD continued to operate east out of Kantara until 1920.

In March 1919 Tim was relieved to be sent to demobilization camp in Kantara. Having boarded the *Mauritania* to take him home, he was abruptly ordered off again to return to duty on the Egyptian State Railway due to the civil unrest in the country. There were strikes and demonstrations demanding independence from Britain, who had more or less invaded Egypt, conscripted over one and a half million Egyptians into the Labour Corps, and requisitioned buildings, crops, and animals for the use of the army. Civil disobedience was met with force and more than 800 Egyptians were killed. Rail tracks were sabotaged and amongst the strikers were the civil rail operators, including station masters and signal men. Tim was sent to Zagazig along with a few other ROD men, but finally managed to get home, arriving in Brightlingsea on 19th May 1919.

Detail from "Palestine Map 6" OS map.
From *The History of The Corps of Royal Engineers Vo; VI.
Chatham. The Institution of Royal Engineers 1952.*

"Turkish loco 3'6" gauge captured at Tul Keram Sept 1918. Hertmann engine."

"At Tul Keram Station. 3'6" gauge locos stock and track soon after the capture by British (Hertmann engine)."

"Runaway train which came to a stand at Tul Keram (Samaria). Coupling broke on steep gradient – no banker – brakes out of order."

"Runaway train come to a stand at Tul Keram Station (Samaria). Draw bar of engine broke and train ran back for about 6 miles down steep decline."

Tul Keram 11.10.1918

Dearest

Thank heaven the sun's going down. Ye gods! It's been a scorcher, and every day is the same, weltering, wiltering, sweltering, scorching sunbeams all day long. Never a cloud, never a minute's shade. And ALL things that creep, crawl, fly, run and walk — God made them all. He surely made them here on the Plains of Sharon, a few have found their way to other parts, but many of every sort are here! The brutes, they bite and sting and poison and torment by night as well as by day. Mosquitoes are the terrors of the night, they inoculate us with fever. And then there are jackals. They go flying about in the night from carcass to carcass and sometimes they seem to pause to ascertain whether you are alive or dead. And these creatures are not all, worse if possible is the heat of the atmosphere. Oh, for a deep breath of the glorious cold fresh air! Breathing and blowing across the Hard at a place so far away called Brightlingsea. I'd freely give a fifty piastre note for one sweet deep breath and think myself well blessed. Thank heaven there's not much work to do.

Quinine, quinine and still more quinine. It's my breakfast dinner and tea. It keeps one's temperature down and makes it possible to get down biscuits and bully. And today — my lucky star — we got jam, real army possie. I have a thirst for iced lemonade like a drunkard has for drink. Everyone has fever in varying degrees. I had it hot and strong three days ago, that was my second day here, but now thank heaven I'm better than many another.

I forget what evening it was that I left good old Jerusalem. A sort of affection had sprung up between myself and the old City. It seemed to have a soul of gold. It was sometimes repulsive, but as Mount Zion and the Tower of David slowly sank from my view I could have shed a silent tear. Swiftly we sank down through the high hills of Judah and suddenly came to a sudden stop — crash! Everybody was thrown from end to end of their trucks — 700 Turks on board had a nasty knock, and so did I. My rifle fell on my forehead and for a while I bled. The medical man who bandaged me sympathised with me for having a aching head but not then or since have I felt anything of it.

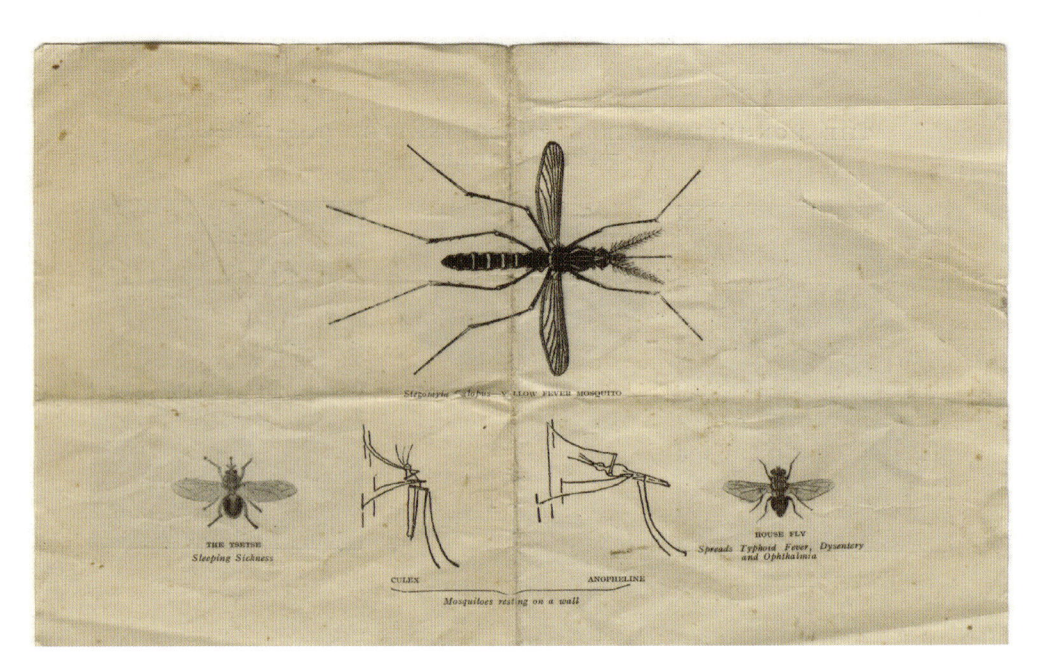

Advert for a "Mosquinette, a protective head-gear, or hood, against all flying insects: for use in Macedonia, Mesopotamia, Palestine, infested tsetse fly African regions, and other tropical and sub-tropical countries."

Poor old Jerusalem! I was and yet I was not sorry to leave it. I wanted a change and I've got it. Up there it is glorious just now — cool breezes blow, the nights are cold and most pleasant, the air is sweet and fresh. Here — on the northern plains of Sharon — we are low down, hemmed in by hills and we are sweating man-sauce and full of fever. But I'm taking care of myself and am more fit than most.

I have yet to answer your last mail. Soon I will do so, Now, the sun's gone, the stars are out and I want to lie...

I hope you are in best of health and spirits — as yet no measle has me molested — A measle! Well I should consider it my mate and take it to the field ambulance and perchance the Doc would let me take it along to Alex where sweet sea breezes flow.

All my love
Your Timothy

Haifa Galilee
Wednesday 21.11.1918

Beloved

The sun has gone down and the moon is just peeping up from behind the Hills of Galilee. I'm on nights this week and at 18.00 (6pm) am on duty. We're getting glorious moonlit nights now, and across the sea it shines as it used to shine across the Colne when we sat by the waterside up the line. I sometimes wonder with the damp, the cold and the rain, I did not kill you in those days. And yet — madman I was — if you did not come I was wretched with the thought you did not want to. But the burning question is, when are we going home? I wonder if my old civvy suit will fit me after 4 long years, and my boots and sox and shirts. Oh, what joy to get into them again, and simply to sit by the old fireside and see in the flames all sorts of wonderful visions. And with you too — you wonderful little woman — looking back upon this my weary exile. I shall not think of myself going through it alone for in everything I feel that you have been with me. I've written and told you all my troubles and you have written lovely letters of sympathy to me. So I shall never recall these dreary days as being borne by me alone dear. Soon I shall be able to tell you all that I have wanted to write. Soon — the joy — the love we have not been able to tell in the touch of a hand — a kiss of the lips or a glance of the eye.

    Again I will wish you a Happy Xmas — where shall we be next Xmas?
    Fondest Love
    Ever Yours
    Timothy

Haifa Galilee
22.11.1918

Dearest

Hurrah! I feel like being mad. A little bit of news has just been posted up in the Mess Room. Good Old Addison — he said in House of Commons that a demobilization scheme had been adopted whereby married men and single men with long service abroad would be sent home first. Also men with situations waiting them would be released early. And demobilization would take place quickly as possible.

God save the King! And may I get home right quick. I'm like a man in a race — standing ready at the starting point waiting for the word GO. And won't I go!! Nothing will stop me. So, just keep your lamps trimmed and burning for you know not the hour etc etc.

I guess you will go all of a heap one of these fine mornings when you see me sitting at the breakfast table when you come down.

Oh you'll have to look slick for I shall want to carry you off like a whirlwind. Why, we got a breeze up now, I'm really excited. I shall want you as soon as I come and I shall want to monopolise you all the time. Get everything done, get the butter made and the ironing done and give all and everyone to understand that you are going on leave — we are to have — oh I don't know how much leave but I believe it's 4 days at least and they are to keep feed and pay us and give us paper money for a spree, and of course you've got to come. Don't let anything stop you — you've done your bit as well as me, and you're to go on the spree with me.

I think I must leave your birthday present until I come home. We will go together and get it.

But do not cease to write to me out here until I tell you — for it must take some time yet before they can get things going to get us home. But look for me among the first from Egypt.

Next Evening

I see it's 28 days, not 14. So just see to it Mrs TCF — we've got a lot of time to make up. You must give them to understand we've got nearly 3 years courting to do and not much time to do it in for we want to marry soon as the gods will let us and we can't afford to forego any of our courting. What are we going to do? In a flash I think of walks round about B'sea, up the line, over the marshes, across the fields. Ipswich, Leondith, theatre, pictures, shops. Sunday evening service at St Mary's Hadleigh, nice quiet and secluded. Oh, I wish we could go to London! But of course I guess we cannot go so far till we are married.

I should like to marry soon as I get home, but of course we shall have to get a home first. I shall have to find a job.

There's a mail in but I cannot get hold of my letters yet. My fingers are itching and my heart seems to have got into my head.

    Ever Your Loving
    Husband

Haifa East Station

Haifa

My Poll

Do not delude yourself that Victoria Place is the only place where your 21st birthday will be celebrated, for here in Galilee it shall be celebrated. It's a long time since I polished my boots and cleaned my buttons, but on the 15th of January 1919 I intend polishing up in honour of the occasion, and when the golden sun has gone west, all by myself I shall stride along the stony streets of Haifa to a certain secluded select little shop that I know of where Abdul Mohammed Hassan sells sweet wine and fragrant fags. After much salaaming I shall bid peace be unto him and desire him depart and not return until he has in his hand a bottle of his best. When he has brought it forth and I am comfortably seated in a cosy corner I shall call upon Allah to give him peace again. Then I shall drink a toast to my wife, Pollie.

Then shall I dream of the days that have been and of the days to be. I shall recall the happy days when we did not care a hang for anyone but each other, when we did our courting before the eyes of all the world between — of all things — school hours, and during business hours too. Down Duke St round into Colne Rd, round again into New St and then through the dirty old gap between Fieldgate and Hibbs, and it's there we would quarrel and occasionally agree and yet we always came again and all the while our hearts were winding and entwining to such an extent that we never realised till recently. In those days I never dared to dream you would ever be my wife, and yet I let my heart drift down upon that stream which it found so pleasing till I actually began to hope, and so here we are. I remember everything. I remember the dear old days. You used to whistle when passing the reading room and if we did not have a happy 15 minutes by the waterside before going to school and work we were wretched and then when school and work were finished, well you used to come and give me a call. Good old Poll. And I would take you for a walk up the Lower Park Rd before tea and then after tea I would pick you up and we would go to the Station together, you to meet Olive, and Oh! Do you remember when first I came to B'sea how I used to meet all you kids coming up from the 5o/c train and the whole blessed lot of them used to giggle and stare much to my embarrassment, but my eye was on you and it was your smile that found its way into my heart. And do you remember when old George Ruffell wanted to know if there was a mouse behind the partition. Oh dear! I nearly died with fear, and the keys I broke! Do you remember all this? Of course you do. I loved you then — I love you now. I used to take you into the first class carriage and kiss you quite a lot. I used to take you up the line too. Oh! I wonder I never killed you.

Just picture me sitting in a cosy corner of the secluded and select shop where Abdul Mohammed Hassan sells rich rare wine in Galilee. And me thinking and dreaming of these things.

Then I shall dream of the days that are to come when my wife and I shall sit and have our supper by the fireside. Milk and biscuits perhaps. We will sip together and break each other's biscuits. I have a faint idea of some sort of a wonderful dress my wife will wear, something soft and warm, something designed for evenings when my wife and I are all on our own by the fireside.

My sweetheart, my little wife, out here in Galilee your birthday — your 21st birthday — shall be celebrated. I shall drink to you, wishing you a happy birthday. Oh! Many happy happier returns. Think of me little woman and if you are not too much taken up with your immediate surroundings just come and sit by my side. I'll make you cosy and comfy and together we will keep up your 21st birthday

Poll Dear

This letter was written before I left Haifa and so it may not be that I shall sit in a cosy corner of the cafe of Abdul Mohammed Hassan and celebrate your birthday. But, Malish (never mind) - wherever I may be the great occasion shall be suitably celebrated by me.

Samakh Railway Station.

Samakh
Syria
Christmas Day 1918

The Day After.

Little Wife

When I had finished writing to you last night I joined in Xmas eve festivities and a high time we had. We sung carols and songs and our doors were open wide to any who perchance were outside. An officer on his way down from Damascus to Egypt was stranded, and he came in and joined us and a real good old camp fire time we had.

This morning is bright and sunny like a sweet spring day in old blighty. Birds are singing, the hills and valleys are green and the rivers and streams are dancing merrily along. Early this morning I ran up into the hills on an engine to do a little business, really in my heart to breathe and feed my soul upon the beauty and sweetness of the morning and the scene.

When I awoke this morning I lay a while and thought of you and wished you, dear, a happy Xmas and a happy happy new year. My fancy naturally wandered on pleasant places, and thinking of marriage I paused a while round our wedding day. I thought I should like when that day dawned to be by your bedside, to come to you while you slept and to gently waken you and to tell you that it was your wedding day. Then to leave you afterwards to have breakfast with the whole family. Soon after breakfast we would go up to the Church, walk, no crush or crowd. You I would have not wreathed in orange blossom with a blessed big veil over your head, no great big bouquet. Maybe early in the morning I would go and get you a buttonhole, say a rose fresh and sweet with morning dew. I should like you to wear something nice and sensible. If it was a dull day, a neat costume, say of serge, or something of that sort, or if it was warm and the sun shone, a nice light dress of silk or something of that sort. No man would ever be prouder of his little wife that I should be of mine, for she would be perfection.

Dear, I must leave you for a little while now. This afternoon I am going shooting again.

Well I went shooting — we were seven rifles strong and about 10 men and Abdul. In the first 2 minutes we had fired 4 shots, four of us. Mine was off a tick or two before the others and got the first duck. A fraction of a second later another fellow fired and got the second duck. After that we did scarcely anything, all the afternoon and only one more duck. We were too big a party and the duck went out over the lake.

In the evening we had a "bust up". Oh, it was some affair! Something after the style of a wild west ranching saloon, and I was boss. 30 thirsty men there were and beer and whisky flowed freely. I had a little whisky in lemonade but not too much. Later in the evening it was a general fiasco. Merriment degenerated into rowdyism and so I set to work dispersing the crowd and soon got them all clear. Then went to bed myself.

But the greatest event of the day was the arrival of a couple of letters posted at B'sea on 4th and 6th December. The mails are improving, but soon the mails will not matter.

Blundell seems to take a great interest in you. I rather fancy he has a soft corner in his heart for you. He came to Liverpool St and saw you off. I guess your mother must have thought him a queer fellow. I'm glad you found a letter waiting you on the mantlepiece when you got home — that's one to me.

You tell me not to get too fat. No fear of that! I guess you will think me too thin. Well, it will be a job for you to do, to fatten me up!

You seem to be having high times in England celebrating the victory. By the time some of us get home you people in England will be sick of the sight of khaki and tired to death hearing about the war. We out here go from day to day now grousing and growling. We've done our bit. We've been out here for 3 years and more without leave, and now the war is over it's not to be wondered that we grouse and growl at our detention in captivity in exile. Let those at home come out and try it — we've had more than enough. I wonder do they think when they give away medals that we shall consider ourselves well rewarded. Give me freedom and they can keep their medals.

Dear, you heave a sigh and say your dancing days are done. Why? What stops you? But perhaps you are squaring your habit to married life. Squaring yourself to a domestic rather than a social life.

"The Sea of Galilee at Samakh on a squally day." (Note railway line running out onto jetty).

Samakh village with the jetty and the railway station to the right.

                    Samakh
                    Syria
            Evening 31.12.18

Dearest

Naturally on the passing of the old new year and the coming new my thoughts should take an extra special turn towards you. I went for a bath this morning in the hot springs. It was great. This afternoon I went for a walk along the shore of the Sea of Galilee. The sun was warm and bright, duck were diving on the lake, beautiful birds — are they the beautiful birds of paradise? Kingfishers, bright and beautiful, dart here and there, diving and catching fish. A solitary little girl I passed, she bade me good day and asked Allah to give peace into me. You would have liked the walk. You would have thought everything wonderful. In the past this part has been out of bounds to civilisation and its beauty but casually referred to in the Bible and not mentioned in modern books nor sang of in songs.

It's raining fast this evening. The water is tumbling down the hillsides and roaring over the rocks down into the river.

We've been issued with a jar of rum on the occasion of New Year's eve — just a wee round, but we're horribly fed up and everyone is in a very rebellious mood, and we all grouse and growl. We get all sorts of rumours about going home but very little official news.

Have many or any of the B'sea boys got home yet?

New Year's Day 1919

Still raining. I was on duty all last night and so saw the old year out and the new year in. During the night you provided me with a lovely hot pudding to eat — the gooseberry pudding you sent me in my parcel. I put it on the fire and it was first class. The night was cold and the pudding was steaming hot and I said what a good boy am I and what a good girl I've got. Can you see me finishing the pudding and saying good old Poll?

Last night I was surprised to find that so persistent had I become in my endeavours to make myself a better man that no room was left for new year's resolutions, so have decided I can do no better than pursue this till little by little and bit by bit I do in fact become better. Just a little bit more than justifies a whole year's persistent effort, and gives courage to go still further. My efforts are most persistent, my strength is greatest, my heart most hopeful and happy — when I get a letter from you.

How did you get on this Xmas and how did you pass from the old year into the new?

I hope you are healthy and happy dear and that the new year will bring you the same happiness. I hope it will bring to myself — that is that wish we used to wish, and which we still wish.

Ever Dear
Your Affectionate Man

Samakh
Syria
5.1.19

Dearest

The days drag along, long, long and weary. They don't go nothing like as quick as they used to. How goes it with you? I guess it's no better. Still, we're nearer to the end than we were and in truth have less cause for discontent than we had before. We haven't had a mail lately and I'm all on the itch for a letter from you. That's very vulgar, but you'll forgive me — it's a treat to be able to write just as one feels, safe with the knowledge that it will be understood. I feel most abominable fed up and I bet you do. We'll grouse together and so get consolation, and when we're tired of grousing, well we'll just cheer each other up! Oh I can be happy with you — grousing or cheering — you're a good old sort.

I dreamt about you the other night. It was a glorious morning. I was standing on a seashore just out of the water fresh and cold. I was in a glorious glow and I saw you in the distance. You waved your hand and laughed and I thought what a glorious world this was and that you were the most glorious girl. Afterwards we had breakfast together, just you and I in a cottage and we were happy. You know what it is dear — you know what it would be just to clasp hands — that alone would be enough to flare into flame the feelings pent up within us. But what would it be to live and love, to live together, to be free, absolutely free?

Oh, Poll dear, sometimes it seems to be beyond endurance, this waiting! I know it's the same with you dear and often I pray hard that I may be free to come to you. Oh, I would fly! And never never again leave you. But of course you know all this, but it seems a bit of a relief to tell it all to you.

But Poll dear we must be patient. We must push along. Never despair, always look ahead. And when things seem hard and the way long and weary, well, we must just hang on and hold fast. And when our time comes we shall appreciate it, by the manner in which we have passed through the shadow so we shall glory in the sunshine.

It's been dull wet and muggy here the last day or so and in such weather the accommodation we have affords us little comfort. It seems funny, here we are right on the Sea of Galilee. It's Sunday evening and away in England maybe our friends now sing of this same Sea of Galilee or else of the Jordan which flows by our side. It seems funny to us here that you should sing of this sea and of this river. I can just fancy myself standing by your side in the Church in Queen St., singing of Jerusalem, Bethlehem, Galilee, Jordan. What memories would be recalled.

Poll dear I believe I've got a bit of the blues. I'm like you, I'm waiting the mail.

Samakh. View Sea of Galilee from the station.

Samakh
Syria
6.1.19

My Dear Old Poll

I cannot possibly sleep tonight until I've let off a little steam. I'm simply boiling over, and if I don't let off steam I shall bust! There's a mail in, and there's 3 or 4 letters for me. Real love letters and Poll dear, I feel happy, glad and jealous. I have come to the conclusion that your letters beat mine into a cocked hat. One time I thought my letters quite passable, I was fairly satisfied with them. But not so now — yours beat mine hollow. I feel jealous of the lovely letters you write, jealous of your ability, proud of your ability too, happy because I am the lucky man. Happy man me. Yet I am sorry that I cannot give you such as you give to me, for I truly feel that my letters now fall far below the level of yours. I am glad too, for I know you don't mind, I know it will give you happiness to put down your hand and pull me up.

Dear old Poll - I cannot conceive how anyone could write love letters to surpass those you write to me, and Poll dear I'm just head over heels in love with you. You know little wife one time I used to calculate that when we married I should have to help you to overcome temperamental difficulties, but really I begin to think the situation is reversed. Poll, you dear old thing, I believe you have been drilling yourself mentally as well as physically.

It's not my intention to attempt to answer your letters now, this letter is to give you something to get on with. You don't want a big gap between letters, you want them often.

Now, excuse me Pollie ~~Pannell~~ Foster but I've lost my balance temporarily. I'm always madly in love with you but manage to keep my equilibrium, but temporarily I've lost it Polly Foster — that's your name isn't it? Oh, by the way, how do you like it? Do you like the Pollie Foster as well as the Pollie Pannell? I think it suits you, but fancy your name being Foster. Fancy people addressing you Mrs Foster, or Mrs TC Foster! It will be a great day Pollie old girl when we sit down to breakfast and we sort out the letters into Mr and Mrs. But I don't think either of us will be so crazy after the postman as we are now. I shall have my wife and it won't matter a lot if the postman hangs himself.

But all this is superfluous, it's the parcel which has upset my equilibrium. <u>Wife dear</u>, it seems so very much as if you are really my wife. Why, my mother is a dear old soul to me! She thinks of every little thing, things for me that I myself forget. She does little things which I notice and which I know no one else would think of unless they loved me as much as she does. Little wife mine you have done everything my mother would do. I was delighted when I opened the parcel, but Poll mine, you must have splashed out and perhaps spent out. Truly Poll I am delighted, and if you spent out you spent out exceedingly well. Oh, it's glorious! You say when buying the socks you really fancied yourself my wife. Equally do I fancy myself your husband in opening the parcel. That parcel Poll is from a wife to a husband - I love to think myself your husband, and this parcel makes me think it. Everything is absolutely it. You must have thought, most intensely, that you were my wife. Everything tells me so. The hankies I wanted them most badly, and you've marked them all, good girl. The socks — they are lovely and civvy — and behold they are marked too, marked exceedingly well! I am delighted Poll, you've forgotten nothing. You've done more than I would have thought of. Poll I can see your love in every little thing. I know you thought me your husband when you were doing all these things. These are not comforts for the troops, they are from a wife with all her love to her husband. You know Poll as I unpacked the parcel I sang praises to the little woman, my wife, who sent it. It's a treat Poll. It seems so much like coming from my wife. It's all from her, nothing but things from her, she chose everything, she did everything, she packed everything. Bless her she's <u>my</u> pet, my good old pal, and it's obvious she must be my wife! It was an excellent idea putting my name on the socks dear.

Only a few minutes before getting the parcel I was asking a fellow to bring me some tooth powder or paste from Haifa as I have none left. Imagine me opening the parcel and finding a tube of tooth paste in it and a tooth brush too. Oh, you don't know what nice things I said about you! And then there's the soap, good old carbolic, priceless treasure — you and the carbolic too. We get Palestine soap issued out here. It's made at Nablus (Shechem) in Samaria of olive oil, and is not up to much. Then there's the "Lemco". Good old Poll, you've thought of me inside and outside. I'm going to lush myself up, or rather you are going to lush me up. But it's the body belt which I place an exceptional value on. I haven't had it long enough to try on but it seems lovely and soft and warm against my face and I know right well

it will be beautiful and comfy when I wear it and that it will keep
me warm. Poll, beloved, you know how I should value it, because it was
yours, that you had worn it, you knew it would all gladden me out here.
It's not the belt alone, it's the love with which you send me with it
and that it has given you warmth. Why old girl, the warmth it gives me
will seem to come from you. Once the warmth it held was yours and now
that warmth is mine. Oh happy happy man! I'm giving special thanks in
my prayers tonight. And then there's chocs and biscuits. I'm full up
wife dear, and there's a towel too... Oh! Oh! I was just going to say
my name should be on the towel and bless me when I look again I find
it's there! Oh you wonderful woman, you're a wife to be proud of and
I'm a happy happy husband.

    Little Wife
    I am Ever
    Your Loving
    Husband

"Captured from the Turks September 1918. Loco at Samakh. 3'6" gauge."

"Loco by A. Borxy Berlin [Borsig 2-8-0]. A link in the Cairo-Constantinople express at Samakh, captured with 3'6" track at Damascus in Sept 1918. At Samakh station Galilee."

[trip to Cairo. No date, first part missing]

... of goods he showed me and tried to sell to me but I told him it was all rubbish and blamed him for wasting my valuable time over such truck. Heart broken, he appealed to his God to come to the rescue and turned to me a piteous face. His prayer was answered for when about to step from his store to the street I turned about told him I liked the look of his face and that in my heart I had found for him a place and that out of pity I would look again at a piece of pale blue calico he hadn't shown me with the rest of the rubbish. Gratitude and love oozed from him as he explained to me that it was the very best quality silk and not calico that he had shown me, and with loving affection he brought forth the said silk and very tenderly fingered and caressed it. Did I want a blouse length or full dress? 2½ yards was necessary for a blouse and 6½ for a full dress. Then followed a pitched battle in which rupees piastres and shillings all took part. Several times I drove the enemy out of his trenches and chased him mercilessly — several times he called upon his God for help but I gained much ground and then turned away disgusted, and told him he was a barefaced rogue and that I should give him so much and no more and the 6½ yards of whatever it may be was torn off and set aside and then he wanted me to buy a lace collar and show me something for which he asked 26/- whereupon I set out to explain to him that I did not want to buy his shop nor even the collar he was handling so lovingly and as I had no horse or dog to wear the stuff he was showing I should be glad if he would show me something at least thinkable for a lady. Caressing his collections of lace collars with tender affection he declared he had nothing better and so after much ado one of his lace collars was placed aside with the pale blue and then the price of the two was added together and once again I turned to go emphatically declaring I could get better stuff for half the price in Cairo. Oh, it was fun! My Hindu companion was tickled to death. At last the price was agreed upon and paid, the parcel done up and Hassoomal Pahloomal out of pure generosity agreed to dispatch it by post for me the very next day. I addressed it and filled up necessary forms and was about to go when a scarf or wrap which he had offered me once again caught my eye. As a parting shot I offered him just half what he asked for it, another battle was fought, another parcel was made, the two wrapped again into one and again I wrote the address. And so you will find two small parcels in one, containing 6½ yards pale blue silk and a lace collar, also an envelope containing those charms I wrote you of so very long ago and a scarf or wrap. The wrap is a cheap and gaudy looking thing I fear, the lace collar is big enough for an elephant. I hadn't the slightest idea as to whether the size mattered. And the length of silk was the only shade of blue I could get. Poll dear I don't know a piece of silk when I see it and so can accept no responsibility as to the quality of the stuff, although I know when you look nice, or rather when what you wear looks pretty. I have no experience whatever in the choice of appropriate or pretty things and so beg to relieve myself of all responsibility in that matter also. Such as they are I send you. Do not look upon anything as a gift not to be parted with, anything you may not care for or do not want, do as you please with, and if in any way you may be pleased even a little then I shall be glad.

What a fearful long rigmarole I've written, but it really was fun. Why we had our money's worth in laughter alone!

Now I'm in my little bed again and your dear sweet face smiles at me from a place quite close to my pillow. I shall be tucked away in my blankets and before sleeping I shall have a few wandering fancies with you, together we will visit our little home. In my mind it's you who shows me these delightful glimpses into our heaven, you who arranges this and that, and takes me from room to room and shows me all the little things which will make us comfy and happy. Many pretty things you show me, for you like pretty things and so do I, but sweeter than all is my little wife.

Darling Good night,
Your Man

29.1.19

Little Wife

I had a rotten ride last night. Left Cairo at 8pm and arrived Luxor at 10am, and by jingo it was cold and I had not my great coat with me. I snoozed and dozed and woke up stiff with cold.

Yesterday I went to the Pyramids and Sphinx. They are great, fairly take the cakes. No one can grasp their magnitude but by seeing them I will send you a snap of myself taken against the Sphinx when it's finished.

I don't know how you would get on if you were with me when we went into the great Pyramid of Cheops. It makes one slip and slide and perspire terrific. The stone passages are like glass, and in places one has to bend low for long distances. In another place one has to go on hands and knees and in other places we sat down on our "seats" to slide down. I all but lost the seat of my slacks through so much sliding and one fellow fainted. You would have to go bare footed or wear rubber soled slippers and be very loosely attired. I guess your "gym" costume would be the thing.

This morning I have been round Luxor temple. This afternoon I go to Karnak temple, and tomorrow I go to the ancient city of Thebes.

The temples and ruins are magnificent, totally dwarfing the antiquity and magnitude of the temples and ruins in Palestine. You could pretty well pass St Paul's or Westminster Abbey through the gateway of the Karnak Temple. Nowhere in the world is there anything so colossal or even as ancient, for these ruins date from 3, 4 and 5 thousands years BC. Compared with these, the temple of Solomon in Jerusalem is modern and yet it has long since passed away.

Now this must bore you, but never mind. I want to tell you of my doings and if it makes you yawn, well, it will not be the first time. Do you remember? You used to yawn and I didn't mind. Neither did you if I have bored you, I've loved you too.

Yesterday I saw in the streets in Cairo a girl exactly like you were when first I saw you but she had no smile.

In haste
I Remain Yours
Tim

Pollie wearing a blue silk dress with a large lace collar.

Sphinx of Giza and pyramid behind

Luxor: Karnak, Colossi of Memnon
and the Mortuary Temple of Ramses II.

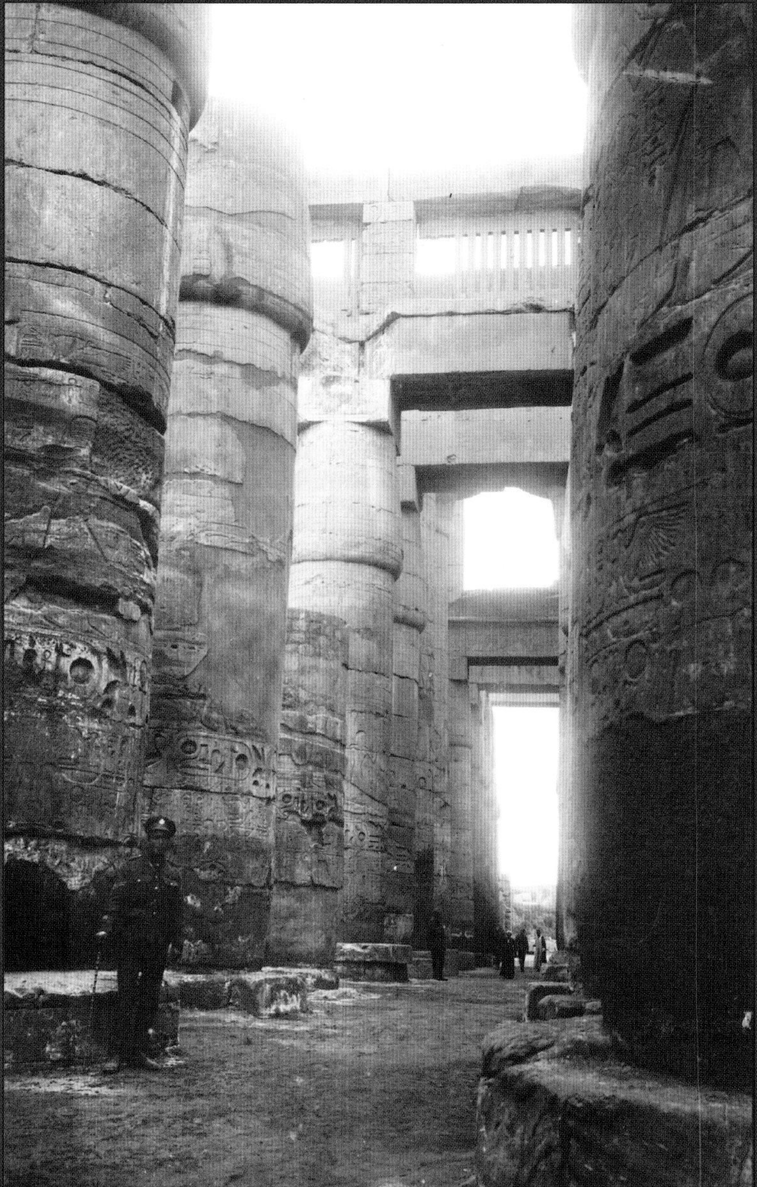
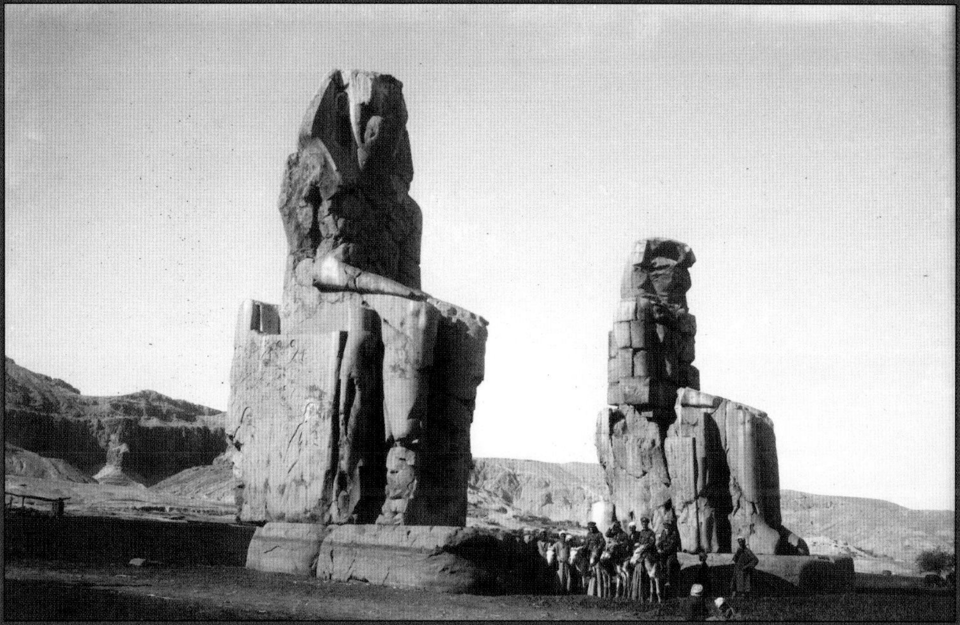
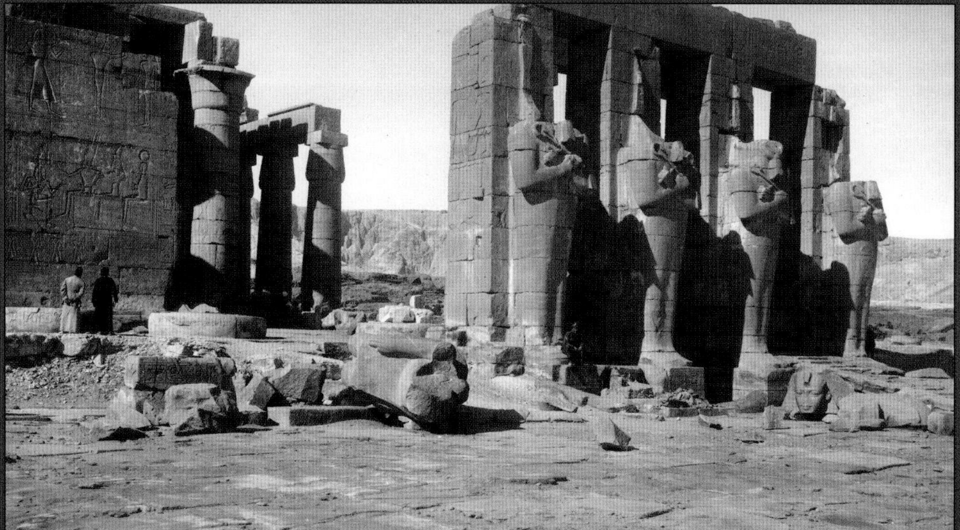

Cairo
Sunday Morning 2.2.19

My Darling

Well! What do you think of my picture? Fred thinks it's like me and so do I, in fact I think it's a very good portrait, but the important point is what do you think of it? You're the judge and I the prisoner at the bar. What do you say your excellency, guilty or not guilty? I'm sending you another print by another post that you may get one should one be lost.

Yesterday morning I went round the Mousky bazaar again. It's a wonderful place. There you can buy and see them selling and grinding precious stones. You can see them preparing spices plentiful and rare too, silk spinning and all manner of silk making, all sorts of laces and other fancy work, iron work, brass and copper work etc etc.

I went and saw old Michael Lonan too see if he had sent off that parcel of silk and stuff to you, and he showed me the insurance policy for it proving its dispatch. The only thing I am doubtful of is if he may have changed the blouse length I chose for something inferior after I left them to pack and send the parcel. Let me know what it's like Poll. I chose the best he had and beat him down badly, and if the old blighter has changed it I will have his life. I rather liked the bit I chose but I'm afraid you may not.

Wandering round the brass workers in search of some small souvenir for Mother and Jess I saw a small brass bowl of engraved Persian work which I rather fancied, and which I thought you might fancy too, and find a suitable place for it in our home. I tried to find a match to go with it and the old fellow said he could not get another for love nor money. So I got the one, and he is sending it with a pair of brass bowls for Mother and a pair of Persian worked bowls for Jess. You can either let it be at Hadleigh till we go, or get Mother to send it on to you. It's only a small thing and was very cheap, but I thought it rather nice. I wish you could have been with me, you would have been delighted with all the wonderful things and you would have known better than I what was really nice and what was not. I had half a mind to buy a beautifully worked brass incense burner. I've seen so much of this sort of thing out here. Do you know what I would do? Well, I know some eleven stone masons up in Jerusalem. I would have got one of them to make me a little stone shrine (pure snow white stone out of King Solomon's Quarry). It would be made for a portrait to be fitted in, and in front on either side would be two small holes for candles to be fitted in, and there in the centre, right before the portrait and

Samakh
Syria
Wednesday 5.2.1919

Little Wife

Well dear, here I am back again in the wilds of Syria among the Arabs and mud huts down in the Jordan valley. It's like being on another globe after Cairo life.

I have a whole budget of letters from you, thank you. I will pull myself together and answer them. Meanwhile I will get off this brief note. It's a bit rotten coming away up here, but so far as the life is concerned I like the country better than the city (with the possible exception of Jerusalem). I'm spent out - stony broke. Still if you want a dollar or two I can let you have it. I've got corn in the mill. I think I will send you another print of my photo with this letter in case the one posted in Cairo did not reach you.

I posted you several letters and PPCs also a parcel from Cairo. I hope they all reach you safely.

Pollie Pannell, I've read "The Blue Lagoon". Parts in particular are very interesting. The nearer one lives to nature the more fully one lives. Much is written in "The Blue Lagoon" but much is left for the reader to read between the lines. I liked the book very much, and was of course particularly interested because I knew that which I was reading and imagining had been read and imagined by you.

Lucky Fred. That's twice or three times he's been home since last I was home. The fellows out here are in a very rebellious mood on this account, and now and again there's a devil of a row. England is asking for troubles and if she doesn't pull herself together and play the game she's doomed. Having done my bit I'm not coming home to slave for 12 hrs a day whilst those who stayed at home live in luxury on the fortunes they made while we were risking our lives for a miserable matter of a few pence a day. This is what all our companions gave their lives for. Is that what I risked my life and endured years of hardship and discomfort for? Maybe I could have a decent job out here or some other foreign land, but is this right? That we who have fought for and saved England should now be denied a decent life or existence in it?

Quite true Poll. There's nothing fair in the army and if England wants to go to war again she will have a job to get men to fight for her. I for one am not fighting again for her. Let those fight who have something to fight for. I'm finished, and the best thing they can do is to let us get home. They can go to blazes all for me, I'm doing no more work.

---

between the two candles and beneath a small archway of pure white stone I would at certain times burn incense in that beautifully worked brass incense burner. Certain days in May, July and August would see the incense winding and rising round and about two steady and ever-burning candles before the picture of the one whom I worship. Not till my life went out would those candles go out. So long as I lived those candles should not cease to burn and would symbolise my love and devotion to the one whom I loved and to whom I was devoted.

This is the last day of my leave. I'm spending the afternoon and evening with Fred - he sends love to you. Tomorrow morning early I start out the way back to Galilee, a long and weary journey. When I get back I shall have covered about 2000 miles in all.

Fondest Love
My Little Wife
Ever Your Loving
Man

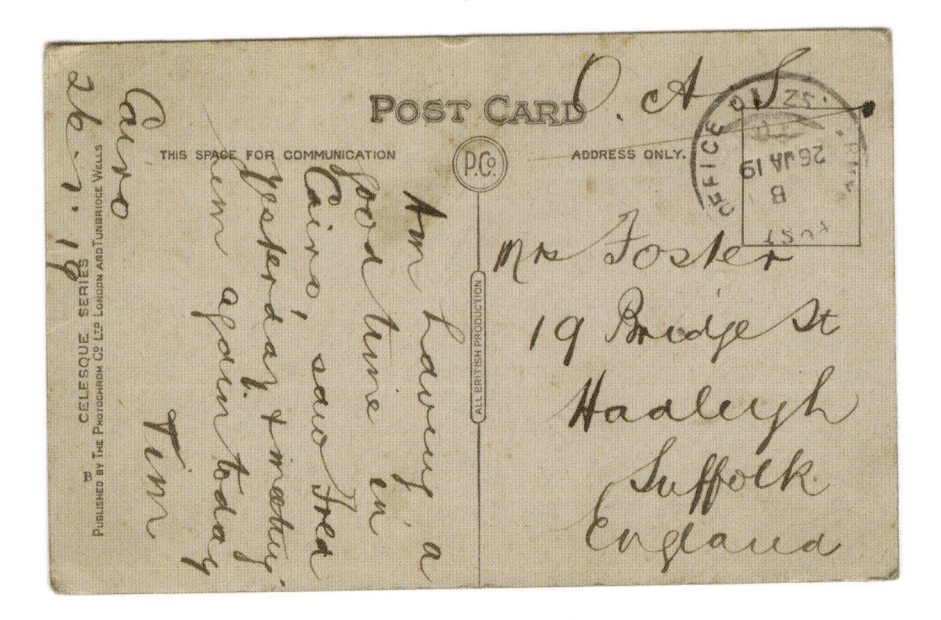

Postcard from Cairo to Mother

Yes, I think you had better be up in the tower when I arrive. They will have quite a job to get us out for some time.

Quite true, you did tell me the ring was to be your birthday present, but look here, I didn't want to be left out in the cold on your 21st birthday. It was only a little thing, but that was enough to give me a look in. As it was they did not send you exactly what I ordered. What they sent was not nearly as nice a thing as I told them to, but still, no matter.

Yes dear, we've got three years' courting to do, and we're going to do it, only we're not taking three years about it. It's got to be done in a few months. We've got to concentrate the very essence of three years' pure joy into a few months. So Poll, just pull your 7st 7 and your 5ft 4 together and look slick. Do you know my little wife that I weigh 10st 11 lbs (I had a half piastre's worth on a weighing machine in Cairo), and my height is 5ft 7 and a bit. And so you can work out for yourself when

Above: Staff at Samakh station.

Left: "Express passenger train 3'6" gauge at Samakh station. A link in the Constantinople-Cairo express which runs from Damascus to Haifa." (After the Yamak bridges had been rebuilt).

I come to you. You're in for a tough time little woman, so just put plenty of pudding away. Your days of killing time are nearly over, you've got to put a little go into it, you're going to be my pal, not a mere play thing and so you've not only to be sweet but you've to be strong too.

I guess it's a great business making your trousseau. I say, it's very interesting to read that you are making a silk nighty. You know dear you've just about stirred up within me as great a liking for pretty things as you yourself possess.

Poll you are my beloved, my wife. I love you dear. I am, all every bit of me, all my life yours. Yours to keep always. You shall love me Poll mine and I shall love you, and I will take care of you. I will make you happy and we shall be divinely happy my darling, that is so long as we go the right way. To work, a certain amount of philosophy is necessary to guide us in the way we should go. It will tell us for instance that one of the most essential factors of happiness is selflessness. We shall we be happy, we shall we go through the years and each find in the other - you in me and I in you - a perfect resting place.

Fondest love, sweetest of all little woman
Forever
I am
Your Sweetheart

Samakh
Wednesday 12.2.19

Darling

Ye heavens, these are exciting times! Iron chains won't hold me beloved once they let me go. I seem to have within me a determined little demon that has a most indomitable spirit that seems to say nothing will stop me. Deep seas and mighty mountains are nothing. I simply laugh at everything and say... I will fly! Oh Poll! Once they let me go I'm coming like a streak of lightning. If you're frightened you can hide, but I shall find you. Sometimes I'm right upon the sky and at other times I'm down in the darkest depths, but in my right senses I reason it cannot be so very long before I am released. They offer good jobs out here, but surely to heaven the country for whom we have fought will find room for us.

I know you don't want to come out here. Besides we do not live to work! We merely work to live. The most important point is not work, not wealth, but happiness. So I make my plans. I seek not work, not wealth, but happiness. Having found that happiness I contrive to keep it and only in so much as work and wealth contributes to that happiness do I give it consideration. How's that my wife? I calculate that if I took a job out here I might be a bit better off, but if in happiness we are a little worse then it's not worth having. We're going to live in England Poll and though we're not going to be wealthy we're going to be happy. What say you dear? I have no letters of yours to answer, we've had no mail the last day or so

I can settle down to nothing. I'm all on the jump to get home.

Samakh
Tuesday 18.2.1919

Dearest

I've got an idea I'm coming home, or at any rate going down to the demobilization camp shortly. Just keep your lamps trimmed and burning for you know not when I may pop in upon you. It may be many weeks, it may be but a few, I believe we go overland via Italy and France.

What's England like? Is it cold and dull and cloudy? And do they still give you ration tickets or books? I hope there's a lot of nots and must-nots and don'ts. It will seem quite strange to again be in a land where roads are hedged and land is divided up and hedged in and one has to keep on the road. Here we wander free and far, no regular road or streets, and not for three years have I been subjected to common civilised customs, manners and courtesies. I fear I shall fare ill among the conventionalities of Blighty. You will have to keep an eye upon me. I'm like a stranger coming to England. You will have to take me round to see the sights and explain to me the manners and customs of those strange folk. Among Bedouins and Arabs maybe I should feel at home but among my own folk I fear I shall fare a stranger.

Goodbye for a while
While
Ever Yours
Timothee

"Hejaz Railway - Girder railway bridge in the Yarmak Valley. Undergoing repairs." (2nd Yarmak bridge).

The 2nd Yarmak bridge after Samakh. "Blown up to cut off retreat of Turks in Sept 1918 and reconstructed by No. 1 Canadian Rly Bridging Co."
Note the fallen box section across the river.

"The Galilean Hills just east of Jordan and Sea of Galilee. At El Hamme Bridge over river Yarmak being reconstructed by Canadian Rly Engineers."
3rd Yarmak bridge. Blown up in September 1918 and reconstruction completed December 1918.

"Hejaz Railway. Rail suspension bridge over the Yarmak, just beyond where it joins the Jordan. Seen though old Roman bridge." (1st Yarmak bridge, before Samakh).

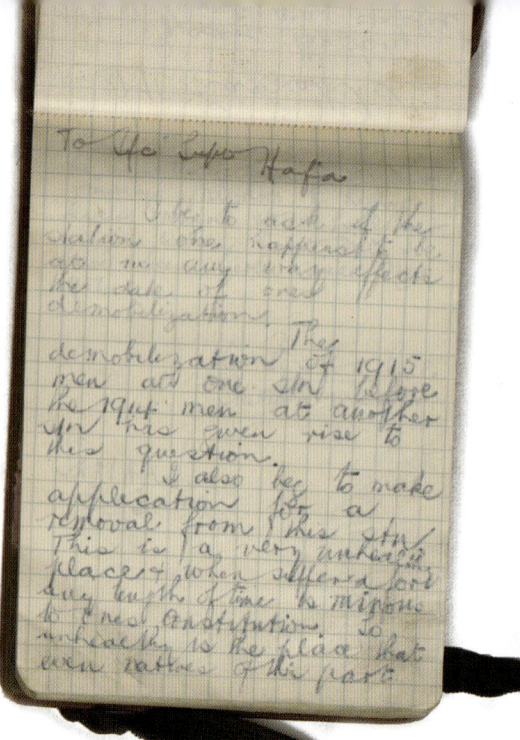

To Traffic Supt, Haifa:

Application for Leave

I beg to ask if you will permit me 14 days leave for the purpose of visiting Damascus.
 Everything has been done to put the civilian employees in the way of working the station and everything is progressing satisfactorily and my duties can now be quite safely arranged.
 I should be glad if this can be granted.

TC Foster Cpl, Station Master
Samakh. 22.2.19

....

To Traffic Supt, Haifa:

I beg to ask if the station one happens to be at in any way affects the date of one's demobilization.
 The demobilization of 1915 men at one station before the 1914 men at another station has given rise to this question.
 I also beg to make application for a removal from this station. This is a very unhealthy place, and when suffered for any length of time is ruinous to one's constitution. So unhealthy is the place that even natives of this part find it unendurable as the weather gets warmer and many migrate to higher levels where it is healthier.
 I would point out I have been here since Nov 1918.

TC Foster Cpl, Station Master
Samakh. 3.3.19

....

MO Samekh:

Bearer No. 3619 Arif Mehemet was crushed between 2 coaches when coupling them together at 1305 Mar 2nd.

He is civilian employee at Samekh Stn.

TC Foster Cpl, Station Master
Samakh. 3.3.19

To Traffic Supt Haifa:

No. 3619 Arif Mehemet

Further to my wire. This man left the hospital before being attended by MO and has been doctoring himself. He has not yet resumed duty but another civilian is performing his duties. The civilian SM has the matter in hand.

TC Foster Cpl, Station Master
Samakh. 4.3.19

To Traffic Supt Haifa:

No. 3619 Arif Mehemet
Civilian Pointsman

Further to my previous correspondence. This man has now been examined by the MO and will probably be fit for duty in a day or so.

TC Foster Cpl, Station Master
Samakh. 5.3.19

To All Concerned:

The bearer Jordan Tocadlidis civilian stationmaster at Samakh has permission to be absent from his duties and proceed to Tiberias for medical attention

TC Foster Cpl, Station Master
Samakh. 5.3.19

To Traffic Supt, Haifa:

Jordan Tocadlidis civilian stationmaster
This man is sick with malaria and will probably be unable to do duty for a day or so.

TC Foster Cpl, Station Master
Samakh. 5.3.19

To OC (Det) ROD RE Haifa:

Soap

For some months we have had no issue of soap although frequent applications have been made for some.
    We therefore cannot be maintained and the health of the staff is impaired. It this place in particular any laxity in sanitation and cleanliness has serious consequences. Will you please send a supply early.

TC Foster Cpl, Station Master
Samakh. 7.3.19

To Traffic Supt Haifa:

Soap

I attach a copy of letter sent to OCL Det ROD RE Haifa today.
    The need for soap is urgent. Men are lousy and clothing cannot be washed and unless soap is sent I fear the consequences will be serious. Already among the civilians malaria and fever is developing.
    Will you please take the matter up.

TC Foster Cpl, Station Master
Samakh. 7.3.19

To Traffic Supt Haifa:

Damascus Leave

I should be glad if you would send the necessary documents for me to proceed by the passenger train on Saty 15th inst.

TC Foster Cpl, Station Master
Samakh. 11.3.19

Samakh
Thursday 13.3.19

Dearest

At last! For over four years I've been waiting for today and now it's come. Tomorrow by the very first train for the demobilization camp at Kantara. Goodbye Galilee. Goodbye Tiberias. Goodbye Jordan. I shall cross it tomorrow for the last time. Thank heaven! This valley is in hot weather a death trap, malaria clasps hold of you and off you go. A day or so ago Ahmed my valet gave me my lunch at 12.30, and at 4 pm I saw him buried. The Jordan valley is one of the most unhealthy places in the world, its graveyards tell the tale. But Cheerio, goodbye forever! I cross the Jordan tomorrow and by the time you get this letter I shall be close upon you. Look out for your 5ft 4 and 7st 7 Poll Pannell.

Ye gods, there's some commotion in Galilee tonight! There's Tahah and Soliman and Jordain and Arif and Yussuf and Mehmet all around. One wants my old bedstead and another wants it too, and all want a lock of my hair. There's the old Admirals from Tiberias tacking here and tacking there across the Sea of Galilee, going to bring me a bottle of wine tomorrow. I can tell you it's some business. I'm quite a notable here and although I am crazy to be up and away to you, I shall softly sigh as the fine sights of Galilee fade from my view, and the old Turks will call upon Allah to bless me and you - for they know of you!

Ships will never sail fast enough nor will trains fly fast enough for me now. Nothing will hold me. In the words of the prophet "Oh for the wings of a dove."

Tonight I recall our early days. I was ever always and all yours before the war began, and right through the war have so remained. I return to you as I left you, this time to remain with you. Poll my pal - this is our day.

All my love
Ever
Your
Man

Kantara
Sunday 16.3.1919
Egypt

Dearest

I guess I've covered something like 300 or 400 miles on the way home. Been travelling ever since 2pm Friday and landed here at 5am this morning. Ye gods! And I am impatient! The blinking train seemed to crawl, and the blessed engine failed right in the middle of the Desert of Sins 100 miles and more from nowhere.

But already I've been through some of the performance - filling up forms, signing papers, kit inspections... Tomorrow morning it's medical inspection etc and in the evening, all being well, off we go to the demob camp there.

At the demob camp we may wait anything from 3 days to 6 weeks meanwhile going through all sorts of formalities.

It's sweltering hot here today. I think the Khamseems are blowing up. You don't know what they are? Great Scott I shall never forget them! Burning hot winds blowing up from over Arabia bringing storms of dust, cruel cruel dust storms, dust storms which stop trains, blinds and buries men. There's no escape, all must suffer it.

I seem to be in a reflective mood with the wilderness behind me and paradise before me. I'm sorry for those of my friends who living and dead I have left away down in the infernal depths of Jordan. I call it the valley of the shadow of death. The living will not be there long, in a week or so men must come up out of it or forever rest in its shadows.

When I look back I have a feeling of thankfulness that I am alive. So many have gone under, even those who worked by my side. So often circumstances have seemed to surround and threaten me with extinction, but still I live and enjoy health better than many. It seems to me I am like a ship which through storms and rocks and wrecks. I am now sailing in sight of the harbour, soon I shall be swinging safely at my mooring, gently rising and falling on the ebb and flow of the tide.

Cheerio Poll. I'm lucky the last train in before I left Samekh brought me what I reckon is my last mail but one from you, and now here at Kantara I have just intercepted what I reckon must be my last letter from you. It was posted on Monday 4th.

There's a bit of rioting going on in Egypt. The people are in a rebellious mood, they want "Home Rule". But I do not think this will effect the day of my sailing. We generally embark at Port Said and travel via Taranto Italy and South of France, or Mediterranean and Marseilles. They say it's a devilish cold journey and takes 6 days by train from Taranto to Cherbourg.

I am glad you received the snaps and like them too. If the films were as good as the prints they would indeed be good. I've got lots more but we will have a look and perhaps print them off when I come home. I guess the great long chap standing by the column is myself, though I'm not so great and long as I look there.

Yes, you shall have another print of my photo. I will enclose it herewith. Am glad you like it so. Jess likes it too but Mother is not struck with it and says I look so much older and my moustache is missing and she likes me best with a moustache.

You wonder if I shall be home by Easter. When is that? I've known no Easter for such a long time. I have no idea when it comes this year.

By Jove it's simply stifling hot here today! Oh for an iced drink! I guess when on my way home I shall sigh for the sun and its warmth. What does one (or two) say when meeting? It's farewell when parting. I cannot say "goodbye parting may be sweet sorrow" but what words can describe meeting, reuniting?!! We say goodbye when parting, but what the dickens shall we say now that we are meeting?

    I am
    On the way to you
    Never to leave you
    Ever
    Your Man

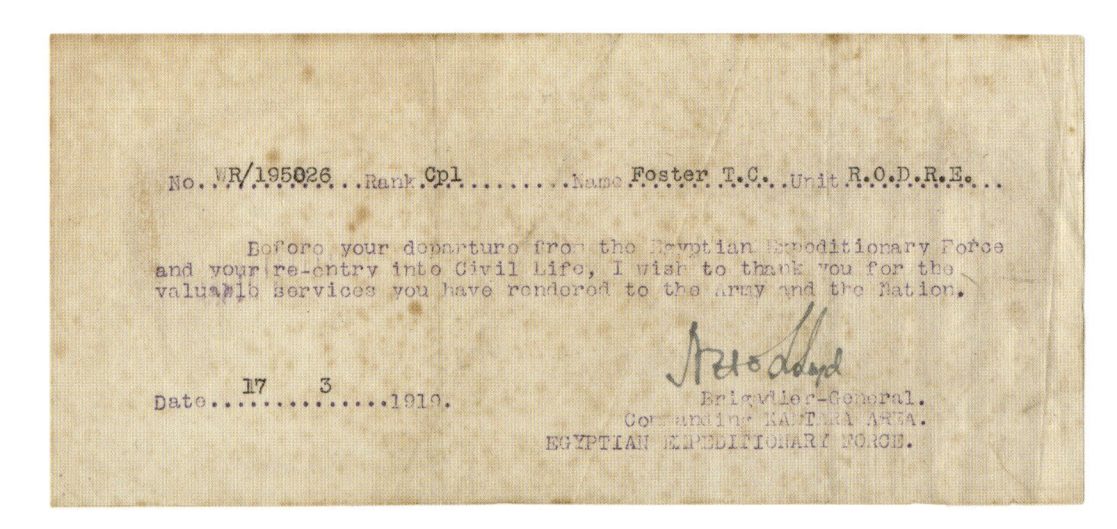

Official thanks for service, 17th March 1919.

Postcard sent to Pollie posted 18th March:
"I'm lucky. 48 hours may see me sailing. It's good old April showers for me."

Spring 1919. The news came from Tim that his turn to return had come at last. I had hoped he would be home for my 21st birthday in January, but the best he could do was to send to a London jeweller's to post me a present. It was a gold brooch in the shape of a wish-bone. March came, and still no definite date for Tim's sailing. I went to the wedding of Olive Aldous and Harold Fry wishing it was my wedding day. The day after, my Granny Pannell died at the age of 86, she was a grand old lady who had looked after herself right to the end of her life.

At last I got a card from him written on the boat he was about to sail in. The next I would hear would be the arrival of the Mauritania in England! The great day came and with it a letter from Tim – not to tell me he had arrived, instead to say he and some others had been taken off the boat to go back on some emergency duty! What a terrible blow! The letters came by afternoon post. As I took it from the postman at the gate, people were coming up the road from the station, one a soldier with kit. It was Frank Bull, one who had at one time been with Tim, he was one of the lucky ones not re-called from the boat. It was the last straw and the biggest blow ever. I went to tell my mother and wept my heart out. My Uncle Fred Clarey was staying with us at the time, he came along and patted me on the shoulder saying "never mind, there are many disappointments as you go through life, he will turn up one of these fine days". This didn't help me much.

I went off back to Moverons, I had come home from there to be ready to go off to Hadleigh when Tim arrived. Needless to say I had a heavy heart as I took my lonely walk through the fields.

I had to possess my soul in patience. I told myself to expect Tim only when I could actually see him.

Kantara Egypt
Monday 25.3.1919

Little Wife

What's your opinion of the British Army and its demob? I should not like to tell you what I think, you would be shocked at my linguistic accomplishments. The fact is Poll, we are delayed. Demobilization for the EEF has temporarily ceased, for how long I cannot say. It may be but a few days it may be for a week or so. There's serious trouble in Egypt. Egyptians have been rioting and Bedouins have risen. People have been killed and a lot of damage done. It may be but a few days and the situation will be in hand and we shall soon sail, but at present it's uncertain. Still I am very hopeful. When the order does come to carry on with demob I shall be in the first draft to go and the order may as I say come any day.

Pollie on the bridge at Hadleigh.

Tuesday 26.3.19

I don't think I shall ever forgive those responsible for holding us up like this. Those who hold us here in exile, who deprive us of our freedom, they retain their freedom, they go home when they like. Are they ever kept from their wives, their families, their friends, so long as we? Have any of them been separated from those they love of 3 or 4 years? Don't they understand? If they would but let us come home for a while, within reasonable durations of absence. No, they will never be forgiven. It's an imposition for which they can give no recompense. Men in England and France who enlisted long after many of us, even long after many of us left England's shores, have long been given their freedom. Is it fair? But for you and those whom I love I would never return. Our prisoners of war are given preference over the British Tommy. We have won! We are the victors, and yet we might as well be the vanquished. German troops have been received home by their own folks and given great welcome, given glory for having fought and done their best. I'm letting off steam - it does me good - don't heed me too seriously, one gets horrible fed up. I was actually on draft and was sailing the next morning, when the news came altering everything. It's all parades just now and huncking one's kit here and huncking it there. No rest - no peace.

I will close and write you again shortly
Keep smiling Little Wife
The clouds will roll by perhaps sooner than we think.

All my love
And Ever
Your Loving
Man

ES Railway
Zagazig Stn
Egypt
2.4.1919

Dearest

I'm particularly pleased with you dear. You've got wisdom and understanding too. You're a real sensible practical little woman and I'm more than proud of you.

I was down - deep down in the darkest depths. Everything seemed black. Letters had ceased to come for I had written some weeks ago stopping them and yet I continued to haunt the ROD Post Office. The news had gone round we were going away into Egypt on the morrow. I was cursing grousing and growling, parading here and parading there, staggering along in the deep sand with about 10 tons of kit and equipment on my back. Flies were biting, mosquitoes were tormenting, the heat and glare of the sun blinded my eyes and made my 10st 11 run - drip drip drip, when bless your dear little heart, I got a letter from you. Didn't I sing your praises! I was happy. That night I lay

"The road to England. The Suez Canal.."

down before sleeping, and in my heart and mind I planned and lived my whole life for you a dozen times over, and laid it down gladly a score or more of times - I made you my god and worshipped you.

There's a lot out here you would like to see.

If a woman loses one she very dearly loves, or a man suffers a similar loss, they regularly go to the place where rests their loved one, and in the evening when darkness is falling they light a candle and sit a while and in undertones talk. They tell their troubles, their hopes and their fears, their joys and their sorrows. Then they come away refreshed, strengthened, and the influence of their loved one has entered into their thoughts and their hearts. So they go and live and act under the same influence as when their loved one lived. So the dead live again. People have their heroes, their gods, to them they go. They worship them and praise them, they receive courage, strength, they tell their troubles, their hopes and fears, and thereby are refreshed and made glad again.

Deep down within my heart I have a shrine - a sacred place - where lives my beloved. There you may find me at any time.

Poll dear - I cannot possibly tell you how I blessed you for that letter.

Before answering it let me just tell you how things stand with me. Yesterday morning we were taken across the Suez Canal and dumped down here at Zagazig in Egypt. A few ROD men are we, standing by to carry on should the Egyptian State Railway men stop work.

But I do not think they will. I think the disturbances are over. The chief thing to do now is to restore the railway lines and telegraph lines, to punish those guilty of damage and murder, and to prevent further disorders.

It is not possible to say when demob will start again but I am of the opinion it will not be above a week or so. When it does we are promised we shall be the first to sail. So you and I cheer. We will make up for all this Poll. Just get a few ideas in your head and we will make things hum when I come home.

Now look here Mrs Timothy, when I come to B'sea, you of all I want to see first. When I've landed in England the first chance I get I shall send you a wire to say I've arrived -

    Miss Poll Pannell
    Brightlingsea
    Just arrived
    Tim

How's that, girl mine? Then I guess it will take pretty well 48 hrs to get through the dispersal camps and home to Hadleigh. One day or evening I will come B'sea first train next morning and get off at Thorrington Station, say somewhere about 10am. Then I will strike out for B'sea and go by way of the stile near Clay's Farm across the fields, and look out for you somewhere before I get to the gate at the corner where stands the village smithy and where runs the back road off the main road. Then of course once together it matters not where we wander - we could go by the back road and across the field and down R Rd or past the school turn and through the fields and down and out by the butcher's and make a bee line across the place de la Victoria. Oh! There's a thousand ways we can wander and get home in time for lunch and take the afternoon or evening train for Hadleigh. I would telephone or wire you my coming the evening before or in the morning so when you get my wire announcing my arrival in England the next thing you get is a message to say I'm coming to B'sea. When I come we leave B'sea again the same day for Hadleigh. Should it be wet or the roads bad, well, I would come to B'sea Station.

Of course if you liked, when you got my wire saying I had arrived in England you could go straight away to Hadleigh and probably arrive there just before me or at same time. It's just as you please.

Now little wife I will read once again your letter. You're a most wonderful girl, you divined precisely the situation, you thought somehow I would not have started - quite correct, and yet I really thought I should have started long ago. And you knew how rotten it would be for me if the letters ceased before I started. I'm glad my plan for spending the first part of the time at Hadleigh is in accordance with your wishes. It must be a dull old hole for you but I think that will suit us all the better for the first part of our time.

I hope this letter may be to you as your letter is to me. I can smile Poll, actually smile when my face is in repose. I had worn a sad scowl before. When enduring disappointments and general depressions we have but little to smile for, and yet by enduring these things we make ourselves better and stronger, and better fitted for the future, better able to appreciate fortune when it smiles upon us.

    All my love
    Little Wife
    Always
    Your
    Man

Zagazig
Saturday 12.4.19

Dearest

Another letter. Good old Poll, you've got what we call "savvy". You're right every time. You know without being told. Like an oasis in the desert your letter is to me more valuable, more precious than all or any precious thing in Egypt. My heart leapt up when I saw it. Poll, you're a brick.

You wonder why you are "blue". I think perhaps it's because I have been. Against England I've got a grudge it will take me a long time to get over. Our legislators are a lot of rogues, their success depending upon their ability to deceive the people. I notice that Mr Churchill stated in the House of Commons that we (the troops who were awaiting demob in Egypt) were appealed to, to stay and defend our fellow countrymen in Egypt. We were not appealed to at all! We had no choice but to stay and remain until they like to let us go. I'm very impatient Poll, and you're the cause of all the trouble. Heaven only knows how much I want to be with you. The longer I know you the more faith I have in you, no one knows you as well as I, and I have that faith in your sympathy and understanding of me that I think you know how impatient I am and how very much I want to be with you.

So you've read of the trouble out here. Well don't worry about me Poll, for I shall take good care of myself, and I really think the worst of the trouble is over now. You tell me not to cuss, but I do, it relieves my feelings a bit. I wish some of those so called statesmen could hear us, it would do them good.

I am very interested to read of the coming wedding of Olive Aldous and a Dinkum. Is she go to Australia with him? And what sort of a fellow is he? I did not think she had sufficient flexibility for a Dinkum, she always seemed so very stiff and straight. Well, I suppose you graced the ceremony with your presence. What was your rig like? I guess you looked very nice and was absolutely *it*. I should like to have seen you. I do not know that I should want to see the bride.

I think you're a plucky kid - you write me very cheerfully and buck me up a lot. I guess I shall have to hang on and cuss (I'll try to grin also) till they like to let me go.

Don't let this letter make you blue old girl, I'll write you a more cheerful one shortly

I am All Yours
Timothy

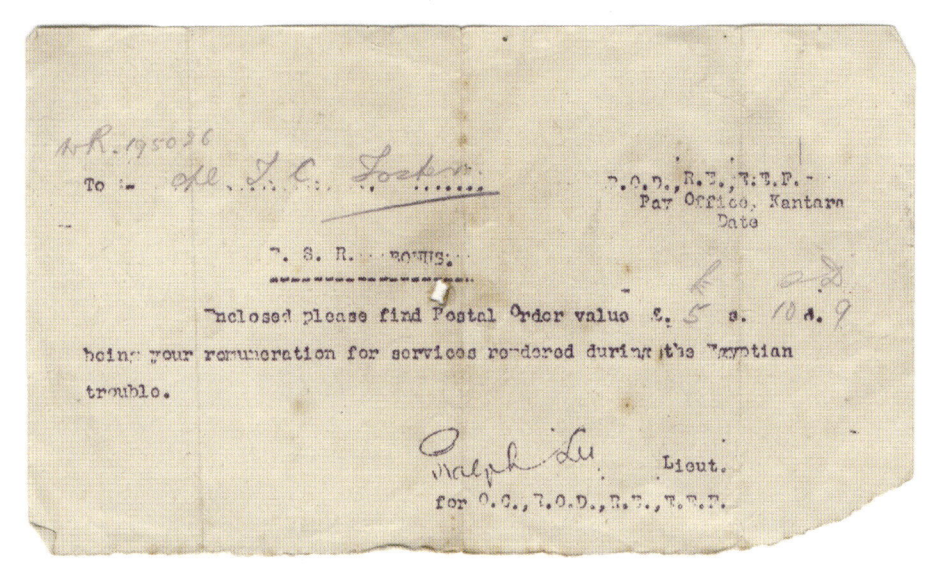

Zagazig Egypt
Easter Sunday
20.4.1919

Little Wife

Our luck's right out just now Poll - Easter Sunday! And only a few weeks ago I would have bet £100 to a penny I would have been with you this day. I can picture Brightlingsea today, a proper old time Easter. I guess you've had my letters and PPCs telling you that I was sailing very shortly etc etc. What a disappointment to you. I don't think I will let you know next time when I expect to sail, there's so many slips between the cup and the lip and it's so bitterly disappointing when the contents of the cup is so cool sweet and refreshing and the lips so dry and hot.

I'm constantly on the alert for any loophole of escape but no luck yet and it's not possible to say when I shall get away. Everything depends upon the situation in Egypt, shipping and reinforcements. At present although the rioting in Egypt is a bit put under, strikes of government employees are more than ever prevalent and the railway men are out in large numbers. The Egyptian State Railways are being run largely by us fellows and we are having to turn our hand to anything. At present I am working a signal box at Zagazig.

A couple of boatloads got away from Egypt this last week but I was not lucky enough to get on one of them.

I should be jolly glad if you would continue to write me. Of course I may get away before, but even so your letters can do no harm. They can easily be returned to the address you put on the back and if I am not away, well, the letters would be a blessing to me. The same old address Poll:
WR195026 etc etc.

I wonder how it's going with you. I have an idea things must be a bit flat and your feelings a bit blue. You have a good bed to sleep in, a table to eat your food from, a chair to sit upon, but I guess all these things must be but part of and contribute to the general flatness and thinness. I calculate you would welcome any diversion or chance to break the monotony and give a little bit of salt to life. Probably the variations I get out here make things easier for me but we both long to have the little excitement of meeting again, of being pals together, of knocking about together, of sharing things, of getting married.

What I should very much like to do would be to get married during my demob leave but there are so many obstacles in the way. One is money. I cannot calculate how much would be necessary to get together such a home as we want. I guess the rough sketch I sent you might be something such as you might be suitable, what do you think? Prices seem to be pretty high in England just now. Then there's the question of getting a suitable house, and where? I think we should be best away from B'sea. Then there's the question of my job. If it's possible to settle all these questions I would say let's marry before I start work. Then we could go for a jolly good honeymoon before settling down and add to the height of the time we will have together when we meet. I should like it much better so, for I'm fed up with being homeless. It's a home I want, and a home of my own. What say you Mrs Timothy?

We won't let little things stop us. All we want is just enough to have a nice comfy pleasant home of our own, and a little bit to spare as a margin wherein we can find room for our friends and be free to follow our own fancies, do as we please, be independent of everyone. Generally to enjoy ourselves and be happy.

I believe these ideas are yours too, but I should very much like to talk things over with you. I fancy that although we could not come up to the home where you now live, you would much rather be Mrs in your own home than Miss where you now are in spite of its many comforts. In spite of all our parents do for us we naturally want to strike out on our own. We love and appreciate our parents none the less. In fact, having launched out on our own we are able to better appreciate and love them for we find out how much they have done for us.

We'll get a hustle on things when we meet, make them hum. How's that trousseau getting on? But look there, we're not waiting for any trousseau once we get started. That's not one of the obstacles. But I've got great faith in you. You've got more savvy than I used to think. Your heart's in the right place, your head's screwed on right, you get a bit blue sometimes, so do I - we've generally got good cause - and you're my best pal, and I'm confident if we knock our heads together we can be ten times happier.

Write me a long letter Poll. I hope I shall not be here to receive it, but that one day when we are no longer interested in the postman it may come back to you endorsed "Gone Home. Return to sender." "Home" may I be and there's only one home for me and that's with you. Without you no house can be by home.

I'm getting a bit sentimental now and when I let myself go that was there's no limit to the leaps of my heart and my pencil is overcome.

Cheerio
Timothee

NO CHARGE FOR DELIVERY

TELEGRAM.

POST OFFICE TELEGRAPHS.

This Form must accompany any inquiry respecting this Telegram.

Office of Origin and Service Instructions.

Charges to pay s. d.

Handed in at

Received here at

Office Stamp.

TO Miss Poll Pannell BRIGHTLINGSEA

Just arrived

Tim

The day came at last, 19th of May, a few days after Fred Lambert was born.

We had arranged by post that we would not meet at B'sea Station in view of the staff – but that Tim should arrive at Thorrington Station where I was to meet him. This plan we carried out! I got to the station with time to spare, and then the train came along with his brown smiling face hanging out of the window – each of us looking four years older than when we parted. At long last he was home! We walked across the fields not meeting a soul until we were well into B'sea. He received a good welcome at Victoria Place.

The next day we went gaily off to Hadleigh to spend his de-mob leave. During that time we also spent a few days at Ipswich with Jess and the children. I felt jealous of the attention Tim gave to his adored sister Jess, even more jealous of the time he spent playing with the children! Blundell cycled over to Hadleigh from Hawstead where he lived then in a derelict old farm house (1/– per week rent!). He lived chiefly on rice and very little else, terribly poor, but happy. He had only been in the army a few weeks, was rejected as unfit.

We had a lovely happy leave, Mother and Father Foster making a great fuss of us, then Tim was at last free of the army. He reported to the railway. At first he worked at Walton Station for a time, then to the District Office at Ipswich. Then he at once spent all his off duty hours house hunting as we were determined to get married as soon as possible. It was pretty hopeless to find a house, flat or rooms: always when he heard of somewhere he was just too late. Tim kept telling me that living on a poor railway clerk's wage was not going to be easy for me, having been brought up in more or less luxury. I didn't care, knowing we would be hard-up didn't worry me. He thought perhaps we should wait until he'd got a better paid job. However we still went on trying for a house, we were eager to have some sort of home. There were disused army huts being let to ex-soldiers, quickly snapped up. I doubt if my parents would have agreed to one of those. Tim put his name down for one, but didn't get a chance, the waiting list was too long, such was the difficulty for the returned soldier.

At last we decided to get married and live at Hadleigh with Tim's parents for a while until we could find a place of our own. This suited my mother much better as she was very depressed and miserable at the prospect of her Babe getting married and leaving home. To know that I would have Tim's mother cheered her a little.

Plans for the wedding went forward, we both badly wanted it to be absolutely quiet with only necessary witnesses (Blundell to be one). In a letter from Palestine Tim had written that when our wedding day came he would first like to call for me early one morning, pick a rose from the garden, no wedding dress and veil, just one of my usual dresses and we would walk together to Church. The wedding to be just for us, no dressed up spectators. My sisters were horrified and said "At least you must have the family!" And so it had to be, not a lovely summer's day as we dreamed about, but a dark dreary stuffy foggy November 17th, 1919.

A few weeks before, I went to Ipswich with Mother and Father to get my dress and other things. First of all choosing the household linen, and by the time we got to the dress department I was getting impatient as it was getting late in the morning and we had arranged that Tim should join us for lunch. He only had an hour I didn't want to be late. I just looked at 2 dresses and decided that the blue (not very pretty) one would do. How foolish I was, could have had the best obtainable. Not until I saw our own daughters married did I wish that I'd had a proper bridal outfit, although I wouldn't have looked as beautiful as they did.

The wedding day was to be on Monday. The Sunday evening before Tim and I went to Church. During the Sermon he slipped my wedding ring on my finger, which I kept on until we said good-night for the last time.

Next morning came, blinding thick yellow fog! Everyone in a flap except myself. My mother miserable because I was leaving home. She couldn't understand how I could be so cool, calm and collected. She couldn't face coming to the Church.

The wedding went off alright. The Church filled up (so much for us wanting it private). During the service I thought my father was going to pass out. I could feel him shaking as he stood beside me. I glanced up at him to see his face ghastly white and trembling.

We rather dreaded the lunch. We managed to stick out the speeches and drinking etc. A few minutes before 12 o'clock we decided we would skip the lunch and catch an earlier train. At 12 I hurriedly said good-bye, Mother sitting sad up the corner with Tom telling her I was happy and not to be so downhearted about me going. Off we went, Blundell deciding he too would miss the lunch, as he too was going to London, so might as well come with us! He was held off at the station from coming in our carriage by those who were seeing us off. Fog signals going off bang as we left the station. Mr and Mrs Tim Foster were off!

Tim and Poll's wedding Day, 17th November 1919, and in 1946.